SpringerBriefs in Electrical and Computer Engineering

For further volumes:
http://www.springer.com/series/10059

S. Srinivasan

Cloud Computing Basics

 Springer

S. Srinivasan
Texas Southern University
Houston
Texas
USA

ISSN 2191-8112   ISSN 2191-8120 (electronic)
ISBN 978-1-4614-7698-6   ISBN 978-1-4614-7699-3 (eBook)
DOI 10.1007/978-1-4614-7699-3
Springer New York Heidelberg Dordrecht London

Library of Congress Control Number: 2014935959

Printed on acid-free paper

Springer is part of Springer Science+Business Media (www.springer.com)

To my family:
Lakshmi, Sowmya, Shankar and Harish

Preface

Cloud Computing has emerged as a cost effective alternative to managing complex computing systems. Traditionally organizations of all sizes required computer networks within their businesses. Managing a network requires specialized expertise which many businesses lack. In many cases network management was outsourced but it still required the attention of businesses since computing service was considered essential. This is where the emergence of Cloud Computing is trying to fill a void where companies would want to focus on their core strengths and let others deal with managing their computing needs. From the start the Cloud Computing model has been attractive because it offers the users the ability to grow incrementally and scale back when services were no longer needed. Moreover, a business could pay only for the services that they use. When there are so many businesses and people who want to use more sophisticated forms of computing services the Cloud Computing model fits well. At the same time it gives a major incentive to cloud providers to offer the necessary services by benefiting from economies of scale.

In this book under the Springer Briefs Series, I will present all the essential details that a potential user would want to consider before deciding on the use of cloud service. The book is divided into seven chapters. In Chapter 1, I will trace the history of the evolution of Cloud Computing as a viable service today and how it is helping entrepreneurs and small businesses have access to advanced services at a fraction of the cost. Chapter 2 will cover the details pertaining to the different forms in which one could use a cloud service. In particular, the reader will be able to understand the three basic types of service that the cloud service provider offers. These services offer varying levels of control to the user regarding their infrastructure. These are known as Software as a Service (SaaS), Platform as a Service (PaaS) and Infrastructure as a Service (IaaS). Besides these three basic types, there are four different deployment modes in which cloud service is used. The most common form of cloud service is known as the Public Cloud. This is more like the Internet where there is nothing proprietary owned by the user. The Private Cloud is often used by large businesses that want to use the architecture of the Public Cloud but keep the computing resources all to themselves without sharing with others. The Private Cloud architecture is usually held in-house as a managed service. It can also be used as a hosted service. There are also two other models of cloud usage—Hybrid Cloud and Community Cloud. The Hybrid cloud comes in two flavors—one in which the user owns a private cloud for much of their computing needs and uses a public cloud primarily for archival purposes. Another way of defining a hybrid cloud is one in which the user uses their computing resources for business critical applications and sensitive data and uses a public cloud for other general uses such as hosting a web site. The Community Cloud is more specialized in that it is used by a group of people with a common interest or a group of businesses that specialize in a particular industry such as automotive, health care or finance. Next I will describe how the cloud is used heavily today for storage and backup.

Chapter 3 discusses the many benefits that Cloud Computing offers as well as the drawbacks associated with the cloud. The details presented here would help a prospective cloud customer as to the things that they should look for in selecting a cloud service provider. Main expectations for the cloud customer are the service availability and the security and privacy of their data. To put things in proper perspective concerning the cloud service I will highlight some of the major outages in cloud service in the recent past so that businesses will be aware of some of the risks associated with a cloud service. Chapter 4 amplifies further on this theme where I describe about the major cloud providers and their services. In this chapter I provide detailed analysis of all the major cloud service providers and niche cloud service providers. In Chapter 5, I address the security aspects related to Cloud Computing. This is an important aspect of Cloud Computing that every user must be concerned with. In this chapter I describe the tools available to protect sensitive data in the cloud. Often businesses are faced with providing evidence to authorities about the compliance requirements such as HIPAA. This chapter will discuss the compliance aspects that a business must consider if it is necessary for them. In this regard I discuss about all the major federal laws in U.S. concerning data and privacy protection that a business may have to comply with as well as industry standards such as the Payment Card Industry Security Standards. Another important aspect that is discussed in this chapter deals with access control that is critical to security.

In Chapter 6, I provide the necessary details that a business would need from a cost and risk perspective in using cloud services. There is detailed discussion on the various types of risks that a business might face when moving their computing resources to the cloud. This discussion sheds light on how using a cloud service is similar to outsourcing. A detailed look at these two aspects shows the similarities and differences between the two types of services. Chapter 7 is the concluding chapter in the book which provides details on what businesses should look for in a Cloud Computing contract in order for them to know how reliable the service would be and how they could limit their liability. Extensive details are presented on what to expect from the cloud service provider and what remedy, if any, the cloud customer would have when a Service Level Agreement provision is violated. The material presented throughout the book focuses on making the cloud customers understand the complexities involved in dealing with the cloud and how they may already be dealing with a cloud service without their knowledge. My goal in this book is to present the material in a simple and easy to understand manner, with several examples highlighting the various topics discussed.

In each chapter I have provided extensive references to validate the points raised and for further exploration of ideas. Many of the web references provide currency of material as of the publication date. Each chapter has Review Questions. This book is aimed at the general business user for them to know the details about cloud computing and also as a learning resource for junior and senior college students in Computer Science and Information Systems programs. The author will be maintaining a companion web site for the book through the publisher which will provide content updates in the blogs. For instructors planning to use this book for classroom use there will be supplemental materials available consisting of PowerPoint Presentations for each Chapter, Question Bank and Answers to Questions. I have carefully checked the validity and accuracy of all the statements in the book. I have carefully proofread all the material in the book. It is very likely that some errors might have escaped my attention. I welcome feedback from the readers on any aspect of the book, including omissions, typos and errors of any type. Please send all communications to the author via email at mvmsrini@yahoo.com.

Houston, Texas   S. Srinivasan, Ph.D.
February 2014

Acknowledgements

I wish to thank my family for the extensive support provided throughout the development of this book. My wife Lakshmi bore the brunt of the work caused by my unavailability for several months while completing this book. Without her unwavering support this book project would not have come to fruition. On the publisher side, Mr. Brett Kurzman has been very patient and encouraging from the start of this project. My deep debt of gratitude goes to Brett for his support. Ms. Rebecca Hytowitz was very helpful in getting the material assembled in proper form and keeping all the communication channels open.

Contents

About the author

S. Srinivasan is the Associate Dean for Academic Affairs and Research as well as a Distinguished Professor in the Jesse H. Jones School of Business at Texas Southern University. Previously he was at the University of Louisville for 23 years building the Information Security program. This program was designated as a National Center of Academic Excellence in Information Assurance Education by the National Security Agency. He received the Ph.D. from the University of Pittsburgh in Pennsylvania. His research is focused on security and privacy. He has published numerous papers in Mathematics and Computer Science as well as presented his research in international, national and regional conferences. Recently he edited a book on Cloud Computing Security. He has actively pursued external grants from federal and state agencies as well as private businesses for many years. He spent his sabbatical leave periods at multinational corporations such as Siemens, UPS and GE. He has volunteered extensively for the profession, community and public education causes. He co-directed the doctoral theses of three students. He has taught Mathematics, Computer Science and Information Systems courses at the undergraduate and graduate levels for over three decades. He serves on the Editorial Boards of some journals. He enjoys walking and reading.

Chapter 1
Cloud Computing Evolution

Abstract Cloud computing has emerged as a cost effective alternative to having reliable computing resources without owning any of the infrastructure. The growth of this technology mirrors the growth of computing in general. The options offered by cloud services fit the needs of businesses of all types. As a truly global technology, cloud computing is growing rapidly, albeit without any global standards. The benefits of cloud computing are too numerous to hold back adoption. At present the goal is to meet the business needs and as the technology matures it will accommodate changes emanating from global standards. As the first step in this direction many of the major cloud service providers are joining multiple consortia to develop the standards. This chapter addresses the history of the growth of cloud computing and the three basic service types—SaaS, PaaS, IaaS—that help businesses of all types. We identify the major cloud service providers and the cloud service types that they offer. We discuss the ways in which cloud computing is supporting entrepreneurial activities. Our analysis shows further that the advancements in communications technology is benefiting cloud computing and makes it a truly global service. Moreover, cloud computing technology is making a major contribution to ecommerce.

Keywords Cloud computing · Technology · Entrepreneurial · Storage · Global · Paradigm shift · Distributed service · Ecommerce

1.1 Introduction

Overall technology growth has been steady. This type of growth spans multiple areas—communications, devices, computing hardware, computing software, video creation, entertainment, etc. One such new technology in computing is Cloud Computing. It is not a revolutionary technology like the Internet was but packages some of the existing technologies in a more user friendly and cost effective way. The success of any new technology is measured by its acceptance by the intended users. As new technologies both telephone and television revolutionized their respective fields. It took telephones over 75 years to gain widespread use. On the other hand, television took only 13 years to gain widespread use. Personal computers which evolved from general computers took only 16 years to become common household item. The internet was launched in 1969 as ARPANET by the US government for use

S. Srinivasan, *Cloud Computing Basics,* SpringerBriefs in Electrical and
Computer Engineering, DOI 10.1007/978-1-4614-7699-3_1,
© Springer Science+Business Media New York 2014

Table 1.1 Summary of technology growth

Technology	Time to gain 50 million users
Telephone	75 years
Radio	38 years
Television	13 years
Personal Computers	16 years
Internet	4 years
Google Search Engine	3 months
YouTube	11 months
Facebook	2 years 10 months
Twitter	3 years

by a select group of research institutions. It grew slowly and gained global acceptance within a span of 4 years once the infrastructure was built. Newer technologies such as the Google Search Engine and YouTube took much shorter to attain critical mass. The Social Media giants Facebook and Twitter started out slow in growth but grew rapidly since then. Today, Google has over 540 million users, Facebook has over 1.3 billion users and Twitter has over 230 million users. Table 1.1 summarizes the time span of acceptance for the old and new technologies.

It is important to realize that the growth of certain technologies depend on the availability of the proper infrastructure. All modern technologies benefit from the availability of several enabling technology. In this sense cloud computing is a beneficiary since the internet is available globally and the communication technology has grown significantly. Because of the availability of higher bandwidth for communication the response time for applications running on a distant server is very low. This feature is known as low latency. Given the availability of these enabling technologies, cloud computing moved the computing infrastructure to the internet. Prior to the advent of cloud computing the server utilization rates were very low. Cloud computing took advantage of the wide availability of internet and the greater communication speed and leveraged the concept of virtual machines (VMs). A virtual machine runs on a physical hardware and since the server utilization rate has been very low, the VM concept was able to launch multiple VMs on a single physical hardware and increase the server utilization rate. This approach is the key to cloud computing where the infrastructure resides with an external provider and serves thousands of customers. Since the customer needs only an internet connection in order to use the cloud computing service, many cloud service providers are able to serve a global audience.

The cloud computing was initiated less than a decade ago. The web based company Amazon, which already had extensive experience running its business over the web, invested heavily in creating the infrastructure that businesses and individuals would need and took the chore of managing a computer system away from the businesses and ordinary users. Because of economies of scale, cloud computing is able to consolidate the services on the cloud and offer the services over the internet. Other companies that had a significant web presence already then launched their own cloud service. Thus grew the cloud services Google Apps, Microsoft Office

365, Windows Azure and Rackspace (Google Apps 2014, Windows Azure 2014). Over the past 5 years, based on customer concerns for privacy protection, the prevailing laws and industry standards many of the cloud services have enacted steps to provide reliable and secure cloud service. These large companies started offering a variety of cloud services that enabled innovation and the launch of several entrepreneurial cloud ventures such as Dropbox, Netflix and Flickr that used the cloud services of these major providers in order to provide their niche services. In the remaining sections of this chapter we will consider the various factors that have contributed to the growth of cloud computing.

1.2   Growth of Technology

Cloud computing today is benefiting from the technological advancements in communication, storage and computing. The basic idea in cloud computing is to take advantage of economies of scale so that IT services could be provided on demand with a decentralized infrastructure. This idea is a natural evolution from the IT timeshare model of the 1960s and 1970s. Today, technology has advanced significantly and many more organizations have computing demands that are elastic in nature. Organizations large and small require reliable computing resources in order to succeed in business. Large businesses deal with complex systems where as Small and Medium sized Enterprises (SMEs) need access to affordable computing resources. Based on these aspects, some of the rationale for today's cloud computing needs can be summarized as follows:

* acquiring and managing the IT resources requires specialized skills
* maintaining a reliable IT infrastructure is expensive
* rapid technology advancements make it difficult to keep current the IT expertise
* internet has opened up many opportunities for individuals as well as small businesses
* number of entities requiring computing resources has grown exponentially
* SMEs' demand for computing resources varies significantly over time
* providing data security is a complex undertaking

In the above paragraph some of the major reasons as to why cloud computing would be advantageous to usehas been identified. The phrase 'cloud computing' is used as a catch all for many types of online computing services. In this book we use the phrase 'cloud computing' in the way most businesses consider—computing over the internet with capability to grow or shrink resources on demand and pay only for the services used. Formally, we follow the definition of NIST developed by Mell and Grance. The NIST definition of 'cloud computing' is "Cloud computing is a model for enabling ubiquitous, convenient, on-demand network access to a shared pool of configurable computing that can be rapidly provisioned and released with minimal management effort or service provider interaction" (Mell and Grance 2011). With this definition of cloud computing we analyze how businesses are using this new

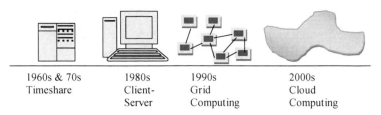

1960s & 70s	1980s	1990s	2000s
Timeshare	Client-	Grid	Cloud
	Server	Computing	Computing

Fig. 1.1 Cloud computing timeline

tool. When a significant part of the business depends on a type of service that the business does not fully control, the question arises as to how the business can meet its obligations to its customers. As highlighted above, IT services are essential to the success of the business but it would be cost prohibitive for many businesses to manage an IT center with the required expertise and fluctuating demand on resources for processing and storage. Thus, a business using cloud computing must understand the security challenges that it would be responsible for and how cloud computing could help in this regard. More details on the security aspects of cloud computing are provided in detail in Chapter 5.

Before looking at the variations of technologies that led up to the concept of 'cloud computing,' let us first understand how this terminology came into use. It is difficult to attribute the first use of the phrase 'cloud computing' to anyone individual. However, it is safe to state that the first known recorded use of the term comes from a 1996 Business Plan by the erstwhile computer company Compaq. Since then Compaq merged with HP (MIT 2011). The phrase is also mentioned in the 1997 Trademark application by a company called NetCentric (Source Digit 2012). The company is now defunct. If any individuals were to be associated with the earliest use of the phrase 'cloud computing' it would be George Favoloro of Compaq and Sean O'Sullivan of NetCentric. Looking back at the evolution of the Internet, it was 1995 when the use of Internet became widespread. Thus, the above timeline of 1996 and 1997 fit well with the introduction of the 'cloud computing' concept. The person who could be credited for popularizing the phrase 'cloud computing' is Eric Schmidt, former CEO of Google. On August 9, 2006, at the Search Engine Strategies Conference in San Jose, CA, while referring to an emerging new computing model he said, "I don't think people have really understood how big this opportunity really is. It starts with the premise that the data services and architecture should be on servers. We call it cloud computing—they should be in a "cloud" somewhere" (Schmidt 2006). This initial talk about this new concept was followed by an actual service called Elastic Compute Cloud (EC2) that was launched by Amazon Web Services on August 24, 2006 (Amazon 2012).

Researching the advancements in technology we are able to come up with the following timeline for the growth of cloud computing. Even though this technology has been around now for over a decade, only now it has become a mainstream technology. We show this timeline in Fig. 1.1 above.

Two of the important facilitators of cloud computing are the Internet and the bandwidth to move large volumes of data. Since the 1960s, cloud computing has grown in many ways even though it was not known by that phrase in the early days. Since the Internet started to offer higher bandwidth in the mid-1990s, benefits of cloud computing for the public at large have evolved rather slowly. One of the major milestones for cloud computing was the launching of Salesforce.com in 1999. It pioneered the concept of delivering enterprise applications via an easy-to-use website. Today salesforce.com is known for its CRM application over the Internet. The next development was Amazon Web Services in 2002, which provided a suite of cloud-based services including storage, computation and even human intelligence through the Amazon Mechanical Turk. Then in 2006, Amazon launched its Elastic Compute Cloud (EC2) as a commercial web service that allows small companies and individuals to lease computers on which to run their own computer applications. Today, Amazon leads in cloud computing services to a vast array of small and medium sized companies and individuals.

The popularity of cloud services relies on the availability of reliable technology to support the service. In this connection the maturity of web in 2009 is noteworthy when Web 2.0 became widely available. Simultaneously, Google Apps provided several browser-based enterprise applications. This is a significant contributor to Software as a Service (SaaS) gaining greater foothold in the marketplace (Google Apps 2014). Besides Google, other companies such as Microsoft, IBM and Rackspace started offering cloud-based services (IDC Report 2012). A significant advantage of cloud-based services is the ability to use virtualization technology to offer each client their own computing infrastructure that appears dedicated but yet shared with other users. Availability of high-speed bandwidth has contributed to the reliability of cloud computing. As far as standards are concerned, they are still evolving with respect to cloud services. In order to reap the full benefits of cloud services the industry needs to adopt universal standards which will greatly enhance interoperability among the various providers.

1.3 A Paradigm Shift in Computing

Cloud computing is a significant shift in the way IT services are managed. Organizations large and small have managed IT services over the years with varying levels of investments. Today, with advancements in communication technology, many new options have opened up for existing businesses and new entrepreneurs want to use more of the capabilities of IT. These two aspects have spawned the significant growth of cloud computing, which gives the customer the ability to benefit from the pay-as-you-go model. Cloud computing has enabled the service providers to benefit from the economies of scale.

This change in service rendering is necessitated by the fact that today's workforce is increasingly mobile and consequently the need for access to remote resources is greater. Moreover, demand fluctuations for IT services are a reality. Businesses cannot afford to provision IT services to meet peak demand, which occur infrequently. Cloud computing provides an ideal solution to meet these needs without incurring significant cost in services provisioning.

Investments necessary to have a reliable IT service kept many prospective entrepreneurs from creating online ventures. On the web, businesses large and small look alike. Cloud computing is providing entrepreneurs the opportunity to try their ideas out, with IT services no longer holding them back as a barrier to entry. The major beneficiaries of cloud computing are small and medium sized businesses as this new concept provides them an opportunity to try out high-end services with no up-front cost, allowing them to use the pay-as-you-go model.

Large enterprises also stand to benefit from cloud computing, although of a different nature. Large enterprises manage data centers and the IT paradigm shift referred to earlier mean more in the context of accessing data from the data centers. In this context private clouds have been introduced where the benefits of storage management and elasticity in demand for computing services are the key drivers. Moreover, the cloud technology also offers high level of reliability and availability of systems without significant capital layout. Often, the benefits of cloud computing are realized by taking a hybrid approach. The hybrid approach gives the large organizations the ability to manage their IT centers and at the same time expand their computing capacity without large capital investment by utilizing the cloud resources. This is especially useful to meet seasonal peak demands with hybrid clouds. Organizations with seasonal high demands that could benefit from hybrid clouds are in the entertainment industry around holidays, sports networks with on-demand service and tax service providers.

In assessing cloud computing's appeal we should consider the usage levels of organizational servers. Server utilization level gives a good metric to see if the investment cost is worth it. The U.S. federal government started looking at the server utilization in its data centers several years ago. It is now widely reported that among all data centers in use the server utilization rate is between 6% and 20%. Even Google's server utilization rate is around 40%. One reason for the low utilization is the lack of virtualization and the need to use dedicated servers for multiple operating systems as well as separation of sensitive applications. Cloud computing is a natural fit to address the low utilization aspect because of higher level of virtualization. With multiple users sharing the computing resources, cloud computing has a very high level of server utilization (Hayes 2008).

Cloud computing architecture enables businesses to meet demand elasticity in computing resources. Business organizations have great difficulty in dealing with demand elasticity for cost considerations. A useful model to compare here is how networks manage elasticity in bandwidth demand. For cost reasons network bandwidth provisioning uses the Committed Information Rate (CIR) model. Likewise, cloud computing provides a similar feature in meeting demand elasticity in both storage and computing power. Without the ability to meet demand elasticity, businesses may end up with an underprovisioned service. In that case customers would abandon such services. Amazon's CEO Jeff Bezos highlights the success of extreme demand for computing power within a very short period of time from Animoto that Amazon was able to accommodate (Bezos 2008). This is a good illustration of high demand elasticity.

In the traditional model, the end user had control over the creation, maintenance and deletion of a document. In the cloud environment, the end user is spared the trouble of maintaining the computing system and reaps the benefits of the application software. This is a positive aspect of cloud storage. However, it is not entirely clear to the end user that when a document is deleted it is going to be inaccessible from the storage system. There have been instances where the document lingered on in the storage system of the cloud provider. These types of issues are unique to cloud computing and thus are a departure from the standard expectation of a computer system. Thus, we note that a shift in approach is needed in order to have control over the online information.

Many large organizations are considering cloud based services as a cost-effective way to plan for disaster recovery. The main cloud service type being considered for this is Infrastructure as a Service (IaaS). Data backup is another service area in which cloud computing is gaining ground. The promise of these two services in the cloud has brought Microsoft and Iron Mountain together to offer data backup and recovery. The customer pays for this service based on the amount of storage used and the retention period for backup data. This service has the added benefit of offsite storage built-in that is essential for disaster recovery and backup because the cloud provider is remotely located relative to the customer. An essential component of efficient data backup is data de-duplication, also known as 'intelligent compression.' The de-duplication method allows for storing only one copy of the data and providing a pointer from all future occurrences of the same data. Data de-duplication can be performed at the file or block level. The latter is more efficient than the former. In typical email backups many users may have the same file as an attachment and so the same file is backed up multiple times. Using the de-duplication approach only one copy is saved and all other references point to the same copy. This is a typical file level de-duplication. Most often de-duplication is more efficient at the block level. In this approach each block of data is hashed using an MD5 or SHA-1 algorithm and the hash index is stored. Future hashes of blocks producing the same hash index are treated as duplicates and not stored. There are sophisticated methods available to detect hash collision, which is rare (Armbrust et al. 2010).

As noted, cloud computing provides a cost effective alternative for users to consider a centralized service that provides computing power, storage alternatives and high reliability. This is a paradigm shift in computing because with the introduction of Client/Server computing more businesses controlled their computing resources and data storage. However, the demand fluctuations for services and the greater need for remote access to data from multiple devices have made it more difficult for many organizations to keep with advancements in technology. This is where we notice a paradigm shift in what cloud computing could provide at a fraction of the cost. Cloud computing's major benefit is the pay-as-you-go business model that it offers. This paradigm shift comes at the cost of security concerns because businesses are hesitant to lose control over where and how their data is stored. The Cloud Security Alliance and the Cloud Industry Forum are providing businesses with many resources to assess the cloud services prior to switching to cloud services (Cloud Security Alliance 2014, Cloud Industry Forum 2014).

Every business is noticing tremendous increase in data availability and the need to store the data. Anytime data is stored it comes with the associated cost of securing as well as backing up the data. This increased demand on the need to make the data available on a continuous basis is making data centers reach their capacity quickly. This forces the businesses to consider options to increase their storage capacity knowing full well that there could be fluctuations in storage needs. The pay-as-you-go model offered by cloud service providers to meet this demand is what makes the cloud an important technology to evaluate. This is a major shift in thinking for many businesses as they realize that they would have to relinquish physical control of their data. This is yet another paradigm shift for large businesses in their use of private clouds.

1.4 Global Marketplace

Many technologies have emerged over the years that have helped facilitate bringing people together in order to grow the world economies. Some of these technologies include various forms of transportation, communication and most recently the Internet. All these technologies had a global impact. Some of these technologies facilitated global commerce, but in localized ways. Internet, however, facilitated the participation of people around the globe in truly enhancing global commerce. In paving the way for global participation, the Internet opened up the ability of entrepreneurs around the world to contribute to global commerce. For example, when one business in a country is seeking a specialized part then the ability to supply that part is no longer limited to that country alone because businesses around the world would be able to know about it and help make that part at an affordable cost. This kind of global competition enables higher quality products to enter the market place and every supplier is aware that in order to compete they have to have an edge over the competition.

In the above paragraph we identified the role of some of the technologies that have contributed to the growth of the global marketplace. Let us now review the relevance of this possibility in the global marketplace with regard to cloud computing. One of the key facilitators of cloud computing is the ability to distribute the information repository in the cloud whereby many entrepreneurs around the world could contribute to the product or service. For example, tax and legal service are two areas where there is a need to have plenty of local information in order to prepare the necessary forms. With the Internet providing the ability to share all the relevant requirements people from around the world with the necessary tax and legal expertise could participate in providing the necessary service at a competitive cost. With the competition now being truly global people will have the opportunity to choose the most reliable and quality service provider. This possibility clearly elevates the quality of service for the consumer. With cloud computing, many entrepreneurs will have the ability to have a web presence at an affordable cost so that they could make their services known to people around the world. Today the publishing world is leveraging this availability of quality providers around the world with the capability to provide timely service. Cloud computing is a major contributor to this success.

Success of cloud computing in the global marketplace depends heavily on the ability to share information across national borders. This information sharing capability relies on the maturity of the ICT (Information and Communication Technologies) sector and the regulatory environment in a country. Fortunately, OECD (Organization for Economic Cooperation and Development) countries have the necessary laws that regulate Intellectual Property (IP) protection. Most of the developed countries around the world are members of the 34-nation OECD countries (OECD 2014). These countries now account for nearly 80 % of the global ICT market that is essential for cloud computing (BSA 2012). Thus, cloud computing is playing a significant role in the global marketplace.

Today, due to the global nature of the Internet, services can be provided from anywhere in the globe. With the widespread availability of cloud computing such providers could customize their services to different parts of the world by using cloud service providers in those respective countries. This is to protect the company's data from trans-border data flow constraints. Countries are realizing the potential benefits of information sharing online and enacting policies accordingly.

1.5   Distributed Nature of Service Provisioning

In order to address the security challenges associated with cloud computing, we need to understand first the meaning of cloud computing. The primary reason for this is that the term 'cloud computing' is used as a catch-all for a wide ranging array of services. After a careful analysis of numerous sources in the literature we have arrived at the following working definition of 'cloud computing' based primarily on the National Institute of Standards and Technology definition: *cloud computing consists of both the infrastructure and services that facilitate reliable on-demand access to resources that can be allocated and released quickly by the user without provider intervention using the pay-as-you-go model* (NIST 2011).

Today's cloud computing has three basic types: Software as a Service (SaaS), Platform as a Service (PaaS) and Infrastructure as a Service (IaaS). In the simplest of terms 'cloud computing' has come to embody SaaS. Similar to the IT time-share model mentioned earlier, SaaS provides both the server hardware and software to an organization without any of the complications of managing an IT system. The simplest example of SaaS service would be email for an organization. The cloud provider benefits from the economies of scale in managing a large infrastructure because of their strength in that area and is able to provide the necessary computing resources to the user, majority of who are SMEs, at an affordable cost. SaaS leaves the full control of the computing system with the provider. Some of the major commercial SaaS providers are Amazon, Google, Microsoft and SalesForce.

SaaS is the most widely used cloud computing service. There are two important sub components in the type of SaaS service available for the users. One type is the Application Service Provider (ASP) model. In this model, the ASP provider is responsible for managing all aspects of the application for its currency and patch management. The user simply pays for the service. The second type is the Hosted Application Management (HAM) model. HAM is somewhat like ASP, except that the HAM provider hosts for the user commercially available software over the web. This is a true software-on-demand application in which the HAM provider hosts a special version of the software created for web distribution. The benefits to the customer are that all end users of the customer use the same software and it gets

Table 1.2 Summary of cloud
service providers

Provider	Type of service	Product name
Amazon	SaaS	AWS
	PaaS	Elastic Beanstalk
	IaaS	EC2
Google	SaaS	Gmail, GoogleDocs
	PaaS	App Engine
Microsoft	PaaS	Azure
Salesforce.com	SaaS	Sales Cloud
	PaaS	Force.com
Rackspace	PaaS	Rackspace Cloud
	IaaS	Rackspace Cloud
IBM	SaaS	CloudBurst
	IaaS	Blue Cloud
EMC	IaaS	Atmos
Apple	SaaS	iCloud
AT & T	SaaS	Synaptic Hosting
VMware	IaaS	vCloud Director

managed the same way for all end users belonging to that customer. We will discuss ASP and HAM relative to cloud services in greater detail in Chap. 2.

PaaS provides the customer a platform, such as the Windows operating system, with the necessary server capacity to run the applications for the customer. The PaaS cloud service provider manages the system for its upkeep and provisioning of tools such as.NET and Java whereas the customer is responsible for the selection of applications that run on the platform of their choice using the available tools. Thus, the customer is responsible for the security challenges associated with the applications that they run. For example, a customer running a SQL Server database on the platform should be aware of the vulnerabilities of the database system. Hence, the customer should have the expertise to manage such applications on the platform used under this pay-as-you-go model. The benefit to the customer is that if their hardware needs change or if they require a Linux/UNIX platform for some other applications, then provisioning them takes only a few days as opposed to few weeks to make the new system operational. One potential drawback with PaaS is that if the customer requires some specialized applications or programming languages, then the customer risks being "locked in" to the provider. Major PaaS cloud service providers are Google App Engine and Windows Azure (Google Apps 2014, Windows Azure 2014).

IaaS provides the customer the same features as PaaS but the customer is fully responsible for the control of the leased infrastructure. IaaS may be viewed as the computing system of the customer that is not owned by them. Unlike PaaS, IaaS requires the organization to have the necessary people with extensive computing expertise. The IaaS customer would be responsible for all security aspects of the system that they use except physical security, which would be handled by the cloud provider. IaaS is also known as Hardware as a Service (HaaS). Amazon and IBM are examples of IaaS providers. Companies with the necessary computing expertise in the form of trained people that wants to test a new service using a different platform would benefit from using IaaS as it would give greater control over the confidentiality of their test. We summarize the results so far in the following table: (Tab. 1.2)

The discussion so far has focused on the three basic types of cloud services—SaaS, PaaS and IaaS. These three types of cloud services aim to meet the customer requirements at different levels of engagement in managing the computing hardware and software. This has a direct correlation to the size of the organization in choosing the type of cloud service. For this reason we can broadly classify the cloud computing users as belonging to either the **public cloud** or the **private cloud**. Small and medium sized businesses typically use the public cloud and large organizations use the private cloud. All the cloud service providers mentioned earlier provide both public and private cloud services. In the private cloud, a large organization which has a data center to manage, is able to use large amounts of storage and computing power dedicated to just their organization only. The private cloud facilitates the large organization to handle demand elasticity similar to the public cloud provider. A natural evolution from these two service models is the **hybrid cloud** which uses both proprietary computing resources and/or private cloud resources that the organization manages directly and the public cloud for some of the computing requirements, especially the ones with varying demands on resources (Bhattacharjee 2009). Two of the major hybrid cloud providers currently are VMware and HP. Another important cloud model that is in use today is the **community cloud**. Community cloud is a form of public cloud that is limited to a particular group having a common interest, a vertical market segment. In such a scenario it would be economical for these organizations to have a dedicated cloud service that caters to the health care industry. For example, health care related organizations such as hospitals, pharmacies, and insurance providers have several things that they have to be compliant with as part of HIPAA (Health Insurance Portability and Accountability Act). A community cloud in this area could focus on the common concerns of the industry partners who belong to this community. The National Institute of Standards and Technology defines a community cloud as "an infrastructure shared by several organizations that supports a specific community that has shared concerns" (NIST 2012). We will discuss in detail these cloud concepts—public, private, hybrid, and community—in Chap. 2.

It is worth noting that the three major types of services described above are gaining ground. According to the Ponemon Institute/CA Technologies 2011 study, among cloud service providers, SaaS accounts for 55%, PaaS accounts for 11% and IaaS accounts for 34% (Ponemon Institute 2011). Another important statistic to note is that 65% provide public cloud service while 18% each provide private cloud and hybrid services.

Earlier in this section we looked at the three basic types of cloud services—SaaS, PaaS and IaaS. Cloud computing supports various types of service provisioning such as availability of software and hardware for use as in SaaS. We will now consider several other ways in which cloud services could be deployed.

Storage as a Service is a business model in which a cloud service provider rents space in their storage infrastructure. Since SaaS is a well-known acronym for Software as a Service, we will use StaaS as the acronym for Storage as a Service. In this model, the StaaS provider will enter into a Service Level Agreement (SLA) with a business needing automated backup and transfer data directly into their storage system. The business will pay for this service both for storage used as well as for the bandwidth used to transfer the data from the business to the storage facility. This service frees up the business to concentrate on their core strengths and the StaaS provider does the data backup for the business. If and when the business needs the stored data, the StaaS provider will be able to transfer the data to the business. Under this agreement the StaaS provider will also be able to provide the added service of encrypting the stored data for the business. This additional service would provide the business the necessary security for the stored data. This service provides an important option for small and medium-sized businesses to promote data backup and plan for disaster recovery and business continuity. Moreover, such businesses will be able to afford having long-term data retention for their important data.

Security as a Service (SecaaS) is an important model that has great potential to succeed. Under this model companies that do not have sufficient expertise in security management can outsource security aspects. Typical applications in this area are deployment of Anti-virus software and network monitoring. This service will benefit more the small and medium sized companies because they lack the expertise needed to manage security for their data. Gartner Research noted that this type of service would generate 60% of revenue in that industry by 2013 (Gartner Research 2013). The major providers of this type of service are Cisco, McAfee, Symantec, Verisign, and Trend Micro. Given the reputation of these companies in the security industry there is a high probability that Security as a Service would be well received.

Today the growth of data available for processing is tremendous. This requires the ability to store large volumes of data and at the same time process the data fast. So, many entrepreneurial companies are developing a new type of service known as *Data as a Service* (DaaS). The basic premise of this model is to generate information in a timely manner by processing a diverse set of data held in text, image, sound and video form. Much of this data is generated by social media. For example, combining position information shared by social network users with time of day gives the ability to advertisers to target market to the right audience their goods. The consumers of this type of service would be both large and small businesses. Some of the successful services based on DaaS are Fidelitone, a 3PL (third party logistics) company; Hoover's, a business data company; Xignite, a financial data company; Urban Mapping, a geographic data company. *Fidelitone* provides logistics solutions to companies specializing in Aftermarket Parts, Electronics, Healthcare, Medical Devices, Consumer Goods, Industrial, and Retail products. *Hoover's*, a Dun & Bradstreet (D&B) company, specializes in making business data available to all users whether they are looking for new leads or a way to keep their CRM up to date. The reputation of D&B, combined with its ability to make available real time data via the cloud, makes this DaaS an indispensable service for many businesses. The main idea behind *Xignite's* business model is that businesses acquire large volumes of data from feeds, databases and EDM systems. Many units in a business could benefit from this financial data if it is processed quickly and made available to them. Xignite apps do just that by storing the data in the cloud once and make the results of the data available to various units by customized apps.Urban Mapping provides geographic information to businesses and using its popular Mapfluence software they are able to provide essential information to a variety of businesses such as health care providers so that they can assess their community penetration.

Testing as a Service (TaaS) is a model that tries to replicate what has been quite successful in industrial applications. Companies that develop a particular product as an offshoot of a larger development process get the product tested by third party companies with expertise in the area. This helps the company to continue to focus on its core strengths and at the same time get a secondary product to market. This type of service is ideally suited for cloud computing when the item being tested is a new software product rather than a physical device. One reason where TaaS companies could be vital in the product testing phase is that in today's rapid development process the product life cycle time is getting shorter constantly. That is where the need for rapid testing becomes critical. In the case of software products, the tester will be able to deploy any type of resource needed in order to complete the testing using cloud computing. TaaS is thus truly an entrepreneurial venture which provides a small business an opportunity to provide an essential service to identify the weaknesses in a proposed service. TaaS also provides the testing company the opportunity to take corrective steps that would help fix the weakness. Thus, the original developer of the service is not only able to get the necessary testing done prior to launch but also have the ability to have a service that will meet the customer expectations. Since the capabilities needed for TaaS are very diverse no one organization could try to invest in all that is needed in order to perform testing and so cloud services become quite indispensable. For this reason TaaS is also known as On-demand testing.Some variations on what TaaS can provide are:

1. Entrepreneurs can serve as consultants to companies on testing because of their specialized knowledge in niche areas
2. TaaS service could be provided without any restriction on geographical location
3. TaaS is well suited for third party testing because the testers need not be fully well versed in all aspects of the product being tested
4. Areas where TaaS could prove useful are in security testing, performance testing, automated testing, and monitoring testing

1.6   Supporting Entrepreneurship

Cloud computing has eliminated the barrier to entry for many entrepreneurs by providing the best of what is needed from an IT perspective. This opportunity has placed an ever increasing demand on having access to large volumes of data. Many datacenters are near their capacity since social media has placed a very heavy burden by generating large volumes of data. This demand has led many data centers to look for cloud computing to provide the necessary infrastructure to handle the data. Google has developed the Map/Reduce technology that facilitates analyzing large volumes of data quickly by chunking a large data set into smaller data sets and processing them in parallel in multiple machines. Apache Software Foundation further enhanced this application by developing the Hadoop framework (Hadoop 2012). These tools, combined with the open source nature of this framework, have spawned several businesses to process large volumes of data on a continuous basis for large businesses. This application shows the need for both public and private clouds. Cloud computing is growing rapidly to support both public clouds and private clouds.

Companies like Amazon, Google and Rackspace support public clouds which are experiencing explosive growth. The data for this growth comes from internet traffic and proliferation of social media applications. The real challenge in meeting future computing demands is in providing connectivity to all types of devices—cell phones, tablets, laptops, desktops and servers. Given the ability to deploy virtual servers, cloud computing is in a better position to meet this challenge. According to an Intel study, by 2015, more than 2.5 billion people with more than 10 billion devices will access the internet, which is more than double that of today's demand (Intel 2010). In order to manage data creation and access by so many devices requires a cloud infrastructure with virtual servers in the order of billion servers. These high demands associated with the data center build-outs are necessary to satisfy the growing demand. Moreover, success in managing such large volumes of data can only be met with the increased efficiency, performance and flexibility of cloud architectures.

Large corporations are facing the need for private clouds because of the expanding business demands on enterprise IT. Many data centers find themselves facing real limits on storage and access bandwidth. When a business considers the option of expanding the infrastructure to meet the growing demand it realizes the many challenges it has to overcome in order to make the investment worth it. Moreover, it is a time consuming process. This is where cloud computing offers an edge because of economies of scale and expertise.

For many companies implementing cloud computing is an evolutionary step because they will have to give up their traditional control over both infrastructure and data. However, it enables the company to realize cost savings along the way in implementing cloud services. It is a fundamental shift in thinking and there are challenges to consider before implementing cloud services. One such instance relates to the company managing itself all their mission critical applications as they transition into the cloud environment. Another important factor to consider is the protection of Intellectual Property (IP) since there is greater potential for loss of data on the cloud. Cloud computing is still emerging and many businesses still consider security as their major concern in using cloud services. Since one of the requirements of cloud services is the ability to automate service provisioning, there should be ways to prevent data leakage in a multi-tenant environment. Newer approaches are still evolving in cloud services to provide the level of automation required and at the same time protect customer data from accidental exposure. This growth period in cloud services should also explore ways to provide flexibility and interoperability for the many applications that are needed by businesses.

Cloud computing's acceptance is growing very rapidly and its economic impact is expected to grow significantly. For example, Software as a Service application services are projected to grow from $ 13.4 billion in 2011 to $ 32.2 billion in 2016 (Market Trends Forecast 2012). Enterprise Cloud applications are projected to grow from $ 22.9 billion in 2011 to $ 67.3 billion in 2016 (IDC 2012). In a related research, Cisco Systems projects that by 2016 global cloud traffic will account for nearly two-thirds of all data center traffic (Cisco 2012). These projections show the promise of cloud computing. Supporting such growth are newer services introduced by Google known as the Google Drive and by Microsoft as the Sky Drive. These services provide free storage in the cloud for individuals and businesses, which could be expanded for a limited fee. Combined with Cisco Systems' observation of more data getting stored in the cloud there is great opportunity looming for entrepreneurs to mine such data for useful information. Facilitating rapid analysis of large volumes of data is also facilitated by new services such as Google Analytics. Our analysis of the existing cloud services and the prospects for future cloud services shows that entrepreneurs have a great opportunity to exploit using the cloud.

1.7 Summary

We have analyzed in this chapter the evolution of cloud computing and traced its origins to the original plan by the erstwhile Compaq Corporation. This analysis is followed by how cloud computing is a paradigm shift and how it relates to the earlier ways of sharing computer resources.Further, we consider briefly the three basic types of cloud services—SaaS, PaaS, IaaS and the four cloud deployment models—public cloud, private cloud, hybrid cloud, community cloud. These concepts are discussed in greater detail in the next chapter. As an extension of this analysis we have included other forms in which cloud computing could help, such as Storage as a Service, Security as a Service, and Testing as a Service. This chapter also includes details on how cloud computing is a truly global technology and how widely it is adopted. We conclude the chapter with a discussion of how cloud computing supports entrepreneurial activities and the financial impact of cloud computing on the global economy.

1.8 Review Questions

1. Give a historical account of the growth of cloud computing.
2. Explain how the concept of cloud computing is a paradigm shift in IT.
3. Explain the technologies that support the global nature of cloud computing.
4. Describe the three basic cloud computing types—SaaS, PaaS and IaaS.
5. Compare and contrast the three cloud computing types and identify the companies that specialize in these basic cloud service types.
6. Describe the four cloud deployment models.

7. Describe four new types of services that take advantage of the cloud and provide specialized service (e.g., security as a service).
8. Explain how cloud computing service supports entrepreneurs.

References

Amazon Web Services. (2012). http://aws.amazon.com/ec2/. Accessed 30 Jan 2014.
Armbrust, M., et al. (2010). A view of cloud computing. *Communications of ACM, 53*(4), 50–58.
Bezos, J. 2008. Animoto, http://www.youtube.com/watch?v=uIc-VB-ke9o Accessed 01/10/2014.
Bhattacharjee, R. (2009). An analysis of the cloud computing platforms, MIT Masters Thesis.
BSA. (2012). Global cloud computing scorecard. http://cloudscorecard.bsa.org/2012/. Accessed 30 Jan 2014.
Cisco Systems. (2012). Cisco Global Cloud Index report.
Cloud Industry Forum. (2014). http://cloudindustryforum.org. Accessed 30 Jan 2014.
Cloud Security Alliance. (2014). https://www.cloudsecurityalliance.org. Accessed 5 Feb 2014.
Gartner Research. (2013). Hype Cycle for 2013. http://www.gartner.com/document/2570118?ref= QuickSearch&sthkw=security%20as%20a%20service. Accessed 1 Oct 2014.
Google Apps. (2014). Google apps for business. http://www.google.com/enterprise/apps/business/. Accessed 5 Feb 2014.
Hadoop. (2012). http://hadoop.apache.org/. Accessed 30 Jan 2014.
Hayes, B. (2008). Cloud computing. *Communications of ACM, 51*(7), 9–11.
IDC Report. (2012). Worldwide SaaS and Cloud Software 2012–2016 forecast.
Intel. (2010). Intel's vision of the ongoing shift to cloud computing. http://software.intel.com/ sites/billboard/sites/default/files/downloads/cloud-computing-intel-cloud-2015-vision.pdf. Accessed 30 Jan 2014.
Market Trends Forecast. (2012). Cloud computing and enterprise software forecast update, 2012. http://www.forbes.com/sites/louiscolumbus/2012/11/08/cloud-computing-and-enterprise-software-forecast-update-2012/. Accessed 5 Feb 2014.
Mell, P. & Grance, T. (2011). The NIST definition of cloud computing. http://csrc.nist.gov/publications/nistpubs/800-145/SP800-145.pdf. Accessed 5 Feb 2014.
MIT. (2011). Who coined 'Cloud Computing?' MIT Technology Review. http://www.idc.com/getdoc.jsp?containerId=238553. Accessed 1 Oct 2014.
NIST. (2012). Cloud computing synopsis and recommendations, SP 800–146, Gaithersburg, MD.
OECD. (2014). http://www.oecd.org. Accessed 30 Jan 2014.
Ponemon Institute. (2011). Security of cloud computing providers study, May.
Schmidt, E. (2006). http://www.google.com/press/podium/ses2006.html. Accessed 30 Jan 2014.
Source Digit. (2012). http://sourcedigit.com/903-netcentric-or-compaq-who-coined-the-term-cloud-computing/. Accessed 30 Jan 2014.
Windows Azure. (2014). The cloud for modern business. http://www.windowsazure.com/en-us/. Accessed 05 Feb 2014.

Chapter 2
Basic Cloud Computing Types

Abstract Cloud computing's marquee feature is the availability of all required software on the web. The principal service that provides this feature is Software as a Service (SaaS) and this is the leading type of service on the cloud. Medium sized businesses that have the ability to have computing expertise amongst its work force have the option of selecting Platform as a Service (PaaS). PaaS gives the business the ability to choose applications that fit their needs most by selecting multiple platforms from the cloud provider. Large and niche businesses have the option of selecting only the infrastructure from the cloud provider, thus benefiting from the Infrastructure as a Service (IaaS) option. In this chapter we will highlight the major benefits and certain drawbacks of these three important services in the cloud. Moreover, the cloud provides different modes in which an organization could benefit. The four basic modes are: public cloud, private cloud, hybrid cloud and community cloud. Public clouds are the most widely deployed and used by all small and medium sized businesses. Private clouds are predominantly used by large businesses that need to supplement their data centers in a reliable way. Hybrid clouds provide a way for a business to manage certain services in-house and use the cloud for some of their customer facing applications. The community cloud serves a vertical market such as health care or automotive where the users have some common features in their applications. This chapter will address the strengths and weaknesses of these four modes of cloud computing. We will conclude this chapter with a survey of Storage as a Service that is gaining ground as an important cloud service.

Keywords SaaS · PaaS · IaaS · Public cloud · Private cloud · Hybrid cloud · Community cloud · Storage

2.1 Introduction

Cloud computing is a global technology that is offering businesses of all types an alternative way to have an information system for their business. Businesses are good at what they do and it is a fact that in today's competitive world they need a reliable computing system to achieve their goals. Traditionally businesses of all types developed their own in-house computing system, with or without help from

S. Srinivasan, *Cloud Computing Basics,* SpringerBriefs in Electrical and Computer Engineering, DOI 10.1007/978-1-4614-7699-3_2, © Springer Science+Business Media New York 2014

external partners. For small and medium sized businesses it is a distraction to have to concentrate on having their computing system functional. It costs both time and money to manage an information system. When cloud computing system became a viable option many businesses decided to use the service. In doing so the business has to give up control of the computing hardware that they will be using as well as their data. This resulted in natural concern both from security and privacy aspects for many businesses.

Cloud computing evolved from the traditional outsourcing model. In this model the necessary service is provided by an organization that specializes in that field. The contracting organization chooses a specific period for which the service is provided. Many small and medium-sized businesses, instead of outsourcing their Information Systems, contracted with specialists to manage their Information Systems. In either case the Information System was under the control of a third party. Often the third party handled all associated data as well. Businesses did not have a problem with this arrangement because of the detailed stipulations spelled out in their customized contract with the third party.

The primary attraction with cloud computing for businesses is the ability to have a fully functional computing system within a few hours or a few days depending on the level of complexity in the chosen system. The cloud computing platform makes available all the options such as the type of hardware needed, the service type, applications needed, and amount of storage, etc. for the customers to select and launch their system (Hurwitz et al. 2012). The access to the system for the customer is through the internet. Cloud customers who need higher level of protection for their communication with the cloud may choose a Virtual Private Network (VPN) connection which is offered by the Internet Service Provider (ISP). The ISP connection speed determines their communication speed with the cloud service. Speeds such as 10 Mbps and higher are quite affordable for many individuals and businesses and so accessing the cloud computing service via their internet service will not be cost prohibitive. Some of the other benefits for a business in using the cloud service are the ability to increase or decrease the use of computing resources, access to a wide variety of applications without any licensing requirement, pay only for the service used, and the ability to have access to high-end computing resources. The most important attraction with the cloud service for a business is that the service is available without the need to maintain and manage the service, which takes significant resource and attention away from the business.

Customers can select from a variety of cloud service providers and so there is better pricing available and the cloud customer could be located anywhere in the world. Many niche services such as payroll processing, human resource management and Customer Relations Management (CRM) that are usually outsourced can be subscribed to as a service through the cloud. Major cloud service providers have invested billions of dollars in their infrastructure with plenty of redundancy built-in for higher reliability. Moreover, the service provider is able to provide a 99.9 % service uptime guarantee. Cloud computing service comes in a variety of service types and deployment models. The most common service types are Software as a Service (SaaS), Platform as a Service (PaaS) and Infrastructure as a Service (IaaS). The most common service deployment models are Public cloud, Private cloud, Hybrid cloud and Community cloud. The major cloud service providers are Amazon Web Services (AWS), Microsoft Office 365 and Windows Azure, Google Apps, Rackspace and Salesforce. AWS is the largest cloud service provider globally and it offers services such as Elastic Compute Cloud (EC2) and Simple Storage Service (S3). Office 365 provides all the office productivity software that many people are accustomed to such as Word, Excel and PowerPoint and the Outlook email service. Azure provides high-end database and search engine services as well as the Skydrive storage service for the consumer and businesses. Google Apps includes all the popular software such as Gmail, Google Docs and Google Drive, plus several more niche services. Rackspace provides all the basic cloud services, including web hosting and the Fanatical Support service for all its cloud offerings. Salesforce is the global leader in offering Customer Relations Management (CRM) software over the cloud. In this chapter we will discuss the details of each service type and each deployment model in detail. We will also introduce additional service types for cloud use that are sought after by businesses.

2.2 SaaS

Today's cloud computing has three basic types: Software as a Service (SaaS), Platform as a Service (PaaS) and Infrastructure as a Service (IaaS). In the simplest of terms 'cloud computing' has come to embody SaaS. Similar to the IT time-share model mentioned in Chap. 1, SaaS provides both the server hardware and software to an organization without any of the complications of managing an IT system. The simplest example of SaaS service would be email for an organization. The first such service was Hotmail from Microsoft in 1996. Prior to Hotmail the email services from providers such as America Online and Compuserve were server based. The cloud provider benefits from the economies of scale in managing a large infrastructure because of their strength in that area and is able to provide the necessary computing resources to the user, majority of who are small and medium sized enterprises (SMEs), at an affordable cost. SaaS leaves the full control of the computing system with the provider. SaaS is also known as the On-Demand software because organizations choose the software that they need from a whole host of software offered by cloud service providers. The early leader in offering SaaS service was IBM in 2003. At that time this service was known as On-Demand software. The term SaaS evolved over a period of time and came into vogue in 2005 when Amazon launched the Elastic Compute Cloud (EC2). Today, some of the major commercial SaaS providers are Amazon, Google, Microsoft and SalesForce.

Managing software such as an office productivity suite of applications requires keeping current on the software, the necessary licenses for all users, patch management and upgrade on the software. Each of these aspects requires the management to devote time and energy. For small and medium sized businesses it is a necessity to have a dependable IT resource but at the same time it detracts them from their core strengths. For example, a small business focusing on manufacturing an automotive part will have to keep up their quality in manufacturing, move the products in the supply chain to the automotive manufacturer and maintain or grow their business. For them, to divert their attention from their core strengths to managing the necessary IT infrastructure would take a toll. This is where SaaS comes handy for small and medium sized businesses. In fact, a service oriented IT department in an organization would immensely benefit from changing their focus from deploying the various software and maintaining them to managing the results of the various applications that SaaS vendors provide.

SaaS evolved as a natural extension on the cloud of the traditional software delivery to businesses. In this early approach, SaaS was essentially a single-tenant service from the service provider. This shifted the maintenance of the software to the cloud provider but did not give the user the benefits of integration from running multiple applications on the cloud. With the growth of technology today SaaS enables the user to work in a multi-tenant environment where the user is able to integrate the results from multiple applications on the cloud.

Businesses see the potential of SaaS as a strategic decision that they have to take to embrace given the risks involved in losing direct control over their applications. In the traditional model vendors sold the software applications to users for a onetime fee and the users were responsible for the upkeep of the software patches provided by the vendor. However, under the SaaS model, the vendor or a third party known as the aggregator provides the software application over the cloud. In this model the user pays an ongoing fee for use of the software on a per user basis without the hassle of maintaining the software. SaaS benefit extends to large enterprises as well. However, in this case the enterprise should have a software management structure whereby all SaaS applications usage in the organization must be cleared by a single group so that the organization could plan for integrating data from all SaaS applications. In many businesses there is a greater need to use both on-premise software and some SaaS applications. Until recently businesses were concerned about the security issues associated with the cloud and there was some resistance to using SaaS applications. With the widespread use of SAS 70 audit features by the cloud service providers many businesses now seem to consider SaaS applications because they will have the necessary audit data to comply with any requirements.

Organizations find another benefit when it comes to using SaaS. Typically the budgeting process in companies requires extensive lead time in order to invest in capital expenditure and also requires significant time for IT to implement the new application. A reasonable estimate of this time frame is 18 months. On the other hand when using SaaS, businesses usually spend less than two months to implement the new application and the budgeting process moves into a different area other than capital expenditure and speeds up the approval process. This ease of implementation also has a potential downside. Many units within a business would find specialized applications that they need and find it in the cloud and deploy quickly. This could lead to mushrooming of applications that a business uses and may have difficulty integrating the outputs from these various applications. That is why businesses should have a strategic vision when it comes to using SaaS as a fast solution to deploy their plan.

Following up on the theme of the previous paragraph, we look at the details of Microsoft Office 365 suite of products. This is a cloud service provided by Microsoft that makes available all their leading software products for office productivity. The main benefit to a business in using Office 365 is that it not only comes with the software that people are accustomed to using in their own computers but it also provides storage for the documents created using Office 365. This feature enables the user to access their documents from anywhere on multiple devices. Today, with the work force being so mobile, access to documents on multiple devices is a must. Moreover, some of these devices use different operating systems, such as iPhone. Instead of the business trying to manage the software on all devices that their employees use and from any location, a business stands to gain in cost and management time from letting the service provided by a cloud provider such as Microsoft using Office 365. Google also provides a similar service using Google Apps. Since many users are already familiar with Microsoft products it helps a business to use Office 365 both from the availability of the necessary software but also providing its work force full access to the documents from anywhere on multiple devices. These are important considerations for a business in evaluating a cloud service.

Typically, in evaluating cloud service businesses consider the pros and cons with regard to the software applications and the hardware maintenance. As mentioned earlier, small and medium sized businesses benefit more from SaaS. The service SaaS consists of a variety of software products that a business uses, not just office productivity software. So, for this reason a business might have to consider other aspects of SaaS when it comes to a multitude of services such as inventory control and Customer Relations Management (CRM). The leader in CRM services on the cloud is Salesforce.com. This being an independent software, businesses will have to evaluate the ease of interoperability between the many software applications that they choose on the cloud. Thus, Office 365 may not suit all businesses. In this regard one software that is worth mentioning is Zoho. Compared to its major rivals Salesforce.com and Google Apps, Zoho provides a free version that could be used by small and medium sized businesses. The approach taken by Zoho is worth mentioning as it offers significant integration of multiple types of applications such as Office products such as email, spreadsheet, inventory management and growing the business using easy tools to follow customer leads. Major software such as Salesforce.com and Google Apps also offer such integration but at a per user cost on a monthly basis which will add up quickly for a small or medium sized business.

Web based services such as SaaS should provide greater flexibility in integrating multiple applications. One such application for many businesses is web based forms through which businesses collect valuable information. Application software such as for CRM should support integration of information gathered through web based services. We emphasized the benefit of SaaS for SMEs in that the business can focus on its core strengths and rent IT services from a cloud provider. Given the need to support many mobile devices for its workforce, SMEs are faced with the

Table 2.1 Interpretation of system uptime metric

System uptime level (%)	Downtime per day	Downtime per month	Downtime per year
99.999	00:00:00.4	00:00:26	00:05:15
99.99	00:00:08	00:04:22	00:52:35
99.9	00:01:26	00:43:49	08:45:56
99	00:14:23	07:18:17	87:39:29

problem of supporting applications on the mobile devices. Companies like Zoho have realized this need and have developed products that will help them remotely troubleshoot mobile devices for specialized applications. Besides Salesforce.com and Zoho other related CRM products are BatchBook and SugarCRM. Each of these software try to provide a niche service such as seamlessly integrating with social media in order to follow customer leads that come through email.

When a business controls its IT the benefit is the knowledge the business gains in knowing the reliability of the system and its uptime. These are two aspects a business must evaluate before deciding on a cloud service provider. In this regard the metric to watch is the system uptime. It is measured as number of 9s, for example, four 9s uptime means that the service provider guarantees that the service would be up 99.99 % of the time. It should be noted that the measure is reported as four nines and not four ninths. The four 9s uptime translates to a total down time of only 52 min a year. Managing a highly available computing infrastructure is very expensive. In order to add one more 9 to the uptime level nearly doubles the cost for the cloud provider. The following table summarizes the meaning of multiple 9s uptime in terms of the respective downtimes per day, per month and per year. It is clear from this table that four 9s uptime provides that a system could be down only 8 s per day. In order to provide such a high level of reliability the cloud service provider must have plenty of redundant systems in place. Companies like Amazon Web Services and Google have the necessary resources to guarantee such a high level of reliability. If a cloud service provider lists in their promotional literature greater reliability then the potential (Table 2.1).

We have mentioned here how a customer could assess the service reliability of a cloud service provider by checking their uptime level. There are simple commands available for the end user to accomplish this. In the Unix/Linux environment the command is $uptime and in the Windows environment the command is systeminfo | find "Up Time". Knowing that such simple tools exist based on the operating system used, the customer need not take the claims of the service provider at face value.

Another major advantage of cloud services is the ability to integrate vertical market applications. For example, in health care or automotive businesses, cloud computing can help eliminate redundancy in the deployment of application software for all businesses in the industry. Using economies of scale, a SaaS provider will be able to provide the necessary common application software to all businesses in the same field such as health care. Likewise, in horizontal market applications such as payroll and customer relations management, cloud computing can help eliminate redundancy in the deployment of application software and their management. It might appear to be lost revenue for the software vendor because they would not be selling their software to many businesses. However, they would still reap the benefits of providing the same service to many more businesses through the cloud. In this case the software vendor would make up the difference in revenue through volume. Moreover, the cloud service would benefit the software vendor in dealing with one provider rather than numerous individual businesses.

One drawback in the use of cloud services either for a vertical or horizontal market is the potential for data leakage. In either type of market the common thread is the application software. When many businesses use the same software for their inventory management or customer relations management they all store data in several virtual servers that share a single data storage device. When a virtual server malfunctions and accesses an area outside of their server then chances are that data would be pulled in a readable format. Another related issue is explained in the following scenario. When data is stored in clouds by multiple banks (all using SaaS from the cloud provider) then the potential exists to search all those data for a particular customer based on the Social Security Number as the key. Why is this important? If in a major investigation the law enforcement wants to find out the financial reach of an individual then cloud facilitates it much easily.

The adoption of SaaS has been rather slow in spite of the many benefits it offers. In the 2008 Forecast survey by Computerworld, many respondents identified several reasons for not considering adoption of SaaS in their operations. The main reasons given were:

1. Security concerns over lack of control
2. Need for enhanced bandwidth to access the application and data over the cloud
3. Lack of offline access to the application
4. Lack of interoperability among multiple applications by different vendors
5. Potential for their data getting comingled with others' data
6. Costly Service Level Agreements (SLAs)

Some of these concerns still persist, especially the interoperability and security aspects.

We described above a general scenario in which cloud services benefit the software applications vendor. We now look at a special case to amplify this command further. Oracle is a major database and Enterprise Resource Planning (ERP) software vendor. They have developed Exadata servers with the specific goal of providing cloud service that combines the benefit of a single provider who is able to provide SaaS applications such as CRM and ERP using the cloud. The benefit Oracle provides in this case is their maturity in both the CRM and ERP markets and their ability to connect these applications using the Exadata servers. Since SaaS applications in the cloud require both fast processors and storage, Oracle's Exadata Database machines are able to meet the demand in terabyte level storage and peta byte level scalability of applications.

2.3   PaaS

Platform as a Service (PaaS) is a cloud based service that gives the subscriber more freedom in the choice of computing platform that they want to use. The PaaS user must have adequate computer specialists to manage the platform that they subscribe to as opposed to a SaaS user. PaaS brings the same level of flexibility that a cloud platform provides with regard to availability of resources and elasticity of demand. Just like SaaS, PaaS also fits the pay-as-you-go model. PaaS provides the customer a platform, such as the Windows operating system with the necessary server capacity to run the applications for the customer. The PaaS cloud service provider manages the system for its upkeep and provisioning of tools such as .NET and Java whereas the customer is responsible for the selection of applications that run on the platform of their choice using the available tools. Thus, the customer is responsible for the security challenges associated with the applications that they run. For example, a customer running a SQL Server database on the platform should be aware of the vulnerabilities of the database system. Hence, the customer should have the expertise to manage such applications on the platform used. The benefit to the customer is that if their hardware needs change or if they require a Linux/UNIX platform for some other applications, then provisioning them takes only a few days as opposed to few weeks to make the new system operational. Major PaaS cloud service providers are Google App Engine, Salesforce.com and Windows Azure.

In PaaS, the cloud service provider makes available several application "components" that the user must put together as needed. This service is somewhat akin to building an object from Lego blocks. For example, Google Apps is a PaaS service from Google where a user could have storage space, ability to work collaboratively with others on textual documents, spreadsheets, presentations and emails. Google Apps is free for those in the education domain whereas it is a paid service for businesses. With over 5 million business users Google Apps is a well known PaaS service. By its very nature PaaS is a complex service that could be managed by organizations that have an experienced IT staff. Because PaaS is available through the cloud it is highly scalable. Even though PaaS gives the user the freedom to choose the applications that run on their platform, the hardware aspect is managed by the cloud provider and so the user can expect to have continuous service with no scheduled downtime for maintenance on a weekly basis. Companies use PaaS to develop and test new applications without the constraints of acquiring the necessary hardware.

PaaS is well suited for large companies and entrepreners for developing, testing and launching new applications based on a variety of platforms. Since the infrastructure cost uses the pay-as-you-go model many entrepreners are able to use a variety of platforms for their applications. Given the ease of use for the end user, applications can be tested in an interactive manner for multiple concurrent users. This kind of load testing is a great benefit to developers. Since resources are all available over the cloud, the developers could create different interfaces for different types of users. Since PaaS users develop their applications on a test platform, testing could

Table 2.2 PaaS features offered by major service providers

Provider	Architecture	ALM	API	Scalable	Log data	Programming languages
Windows Azure	Multi-tenant	Yes	.NET and REST	Yes	Yes	C#
EMC Private Cloud	Multi-tenant	N/A	Atmos	Yes	Yes	Java
Amazon Elastic Cloud	Multi-tenant	Yes	Proprietary	Yes	Yes	C++, C#, Java, Perl, Python, Ruby
Red Hat Open Shift	Special software layer	Yes	Support all major cloud service providers	N/A	Available through provider	Java, Ruby
Ubuntu Private Cloud	Open Source	N/A	Based on Amazon's	N/A	Yes	N/A

involve sharing applications by multiple users while planning for scalability and security. Another benefit of PaaS is that it allows the developer to form distributed teams that work concurrently on various aspects of their application and assign different users different levels of access and track their usage patterns during the testing phase. Thus, PaaS offers a flexible multi-tenant architecture.

One of the key benefits of PaaS is that it supports the complete life cycle of any application development. This process would involve providing features that the customer should be able to combine in the ways they want and create the necessary applications. Such users need not be traditional programmers, rather typical users in a practical environment. PaaS features support the ability to collect logs of user patterns and identify any problems that occur when a real user tries out a new application. It is important for a developer to know that the platform supports Application Lifecycle Management (ALM) since it is essential for the developer to know that future changes would be easy to implement. With this feature of PaaS in mind we recommend that a potential user should evaluate the following aspects before deciding on a particular provider:

- Does the platform support multi-tenancy in architecture and applications?
- What Application Lifecycle Management applications are supported?
- What Application Programming Interfaces (APIs) are supported?
- Does the platform facilitate scalability?
- What types of log data would be available for the user?
- What programming languages are supported by the platforms?

We conclude this section with the following table that summarizes the availability of these features with respect to some of the major PaaS services. One new concept referred to in the table below refers to REST (REpresentational State Transfer) which uses HTTP methods explicitly and is stateless. REST was introduced by Roy Fielding in his doctoral thesis at the University of California-Irvine in 2000. REST is widely used today in web services. REST uses four HTTP methods: GET,

PUT, POST, and DELETE. REST is an alternative to SOAP (Simple Object Access Protocol) based web services. For example, when the browser makes a request for GET or PUT the response could be a PDF file, an image or an XML output. For this reason REST is preferred in web services because it preserves the state of what is being transferred to the recipient (Fielding 2000) (Table 2.2).

2.4   IaaS

Infrastructure as a Service (IaaS) provides the customer the same features as PaaS but the customer is fully responsible for the control of the leased infrastructure. IaaS may be viewed as the computing system of the customer that is not owned by them (Combs 2012). Unlike PaaS, IaaS requires the organization to have the necessary people with extensive computing expertise. IaaS is also known as "utility computing" since the organization needs the computing resources but does not invest in it directly but acquires the resources just like it would acquire a utility such as electricity and water. The IaaS customer would be responsible for all security aspects of the system that they use except physical security, which would be handled by the cloud provider. Amazon, Rackspace, Xerox and IBM are examples of IaaS providers. Typical use for IaaS is when a developer builds an application on a virtual machine of the cloud service provider and customizes the application to the needs of various customers by running them on multiple virtual servers. In this case the large organization is able to take advantage of the availability of virtual machines and manages the VMs for running their specialized applications. Of the three services IaaS is the most expensive and it is used by large corporations. The use of IaaS could be a supplement to the in-house computing resources of the organization. As described above, IaaS could also be deployed for certain applications using a VM environment. We summarize in Table 2.3 the major providers of these three types of cloud services.

The major strength of IaaS cloud service is that it extends the capabilities of large organizations in enhancing their IT resources. Many organizations are able to modernize their IT infrastructure using IaaS without the capital outlay needed to expand their corporate IT. Since organizations pay only for the cloud resources they use, the IaaS architecture provides the traditional benefits of the cloud and yet gives the customer more control over the security aspects of the applications that run on their virtualized environment. The scope of IaaS is many-fold. Organizations could use IaaS for pure computing power, which would be hardware and software. IaaS is also used by organizations for a specific purpose such as storage, security, or networking (Gartner Magic Quadrant 2013).

The real benefit to large organizations using IaaS is in acquiring raw computing power without the capital outlay. Since the organization controls what the infrastructure is used for it has the best of both worlds—access to computing resources as well as control over the infrastructure. The main reason for acquiring such additional computing power is to find ways to integrate multiple applications. For exam-

Table 2.3 Summary of cloud service providers. (Source: Srinivasan 2014)

Provider	Type of service	Product name
Amazon	SaaS	AWS
	PaaS	Elastic Beanstalk
	IaaS	EC2, S3
Google	SaaS	Gmail, GoogleDocs
	PaaS	App Engine
Microsoft	PaaS	Azure
Salesforce.com	SaaS	Sales Cloud
	PaaS	Force.com
Rackspace	PaaS	Rackspace Cloud
	IaaS	Rackspace Cloud
IBM	SaaS	CloudBurst
	IaaS	Blue Cloud
EMC	IaaS	Atmos
Apple	SaaS	iCloud
AT & T	SaaS	Synaptic Hosting
VMware	IaaS	vCloud Director

ple, when the CRM application runs on SalesForce.com web site, the hardware the organization needs to integrate a part of the output from SalesForce.com to another application running in another cloud service provider may not be available to the organization. These applications need not be IaaS as well. They could be PaaS or SaaS applications. Thus, the organization will have plenty of need to test many scenarios of integration and it might need additional hardware unhindered by corporate security policies that might prohibit moving data across platforms. The IaaS provides the necessary infrastructure under the control of the organization with respect to its operation. In a similar way, the organization might have temporary needs to store large volumes of data for retention purposes for a specific period. Once again, the control the organization has in managing the external storage is what makes it acceptable for the organization. In this case the organization has to look at the nature of storage—is it shared in a multi-tenant environment or is it dedicated to a single user. The latter type of storage is more expensive but yet it gives the organization the necessary data concerning compliance requirements. The one drawback in this scenario for storage as a service using IaaS is that the organization becomes responsible for backup and recovery of any stored data. If the organization, on the other hand, opts for a third party storage as a service then the backup and recovery are taken care of by the cloud service provider.

We listed above that security and networking are also aspects under which an organization could use IaaS. The most challenging part in using IaaS would be with respect to security. Usually security aspects require full control over the infrastructure. With the use of IaaS for security, the organization will have to cede the physical security of the infrastructure to the cloud service provider. This is not a major aspect and since the cloud service providers have a reputation to uphold they would be careful to provide the necessary physical security for the infrastructure. The organization should then be able to use their personnel to device the necessary

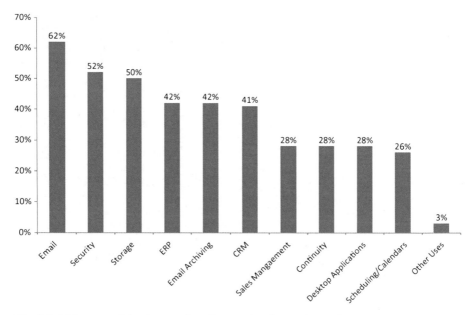

Fig. 2.1 Major uses of cloud computing. (Source: Loudhouse Research)

access security policies and their implementation. The cloud provides the ability to monitor many aspects of the service and thus have more data points to evaluate security. On the networking side, the true benefit from IaaS comes with the ability to deploy a variety of hardware from multiple locations to connect the various segments of the organization spread over a large geographical area. Since the service is a pay-as-you-go model, the ability to deploy the necessary infrastructure is facilitated by IaaS.

It is worth noting that many organizations find the IaaS service as instrumental in reducing the pressure on their infrastructure. In a survey conducted by Loudhouse Research in USA and UK of IT decision makers found a very high level of satisfaction among the users for the cloud service that they were getting. One key finding of this survey was that 73 % of the users reduced their infrastructure costs because of the availability of cloud services using IaaS. This survey further reinforced other beliefs about the use of cloud services (Loudhouse Research 2010). The specific areas here are:

- Lowered the cost of IT use 57 %
- Better control of data 58 %
- Overall better organizational performance 61 %
- Increased end user experience 72 %

Some additional information gained from this survey of 502 IT decision makers reveals the following. The significance of the survey results is that it was conducted by Mimecast through personal interviews of the decision makers (Fig. 2.1).

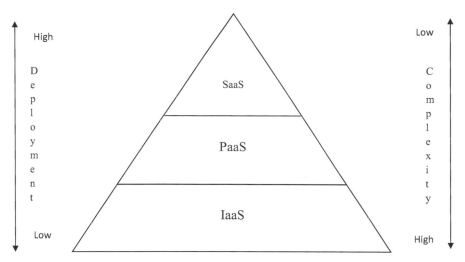

Fig. 2.2 Comparative view of basic cloud service types

We conclude this section with a consolidated view of these three types of cloud services and their level of deployment in the market place along with their complexity (Fig. 2.2).

2.5   Public Cloud

The three types of cloud services discussed in the previous sections aim to meet the customer requirements at different levels of engagement in managing the computing hardware and software. This has a direct correlation to the size of the organization in choosing the type of cloud service. Related to the service types are the service deployment models. These deployment models can be classified as Public Cloud, Private Cloud, Hybrid Cloud and Community Cloud. In this section we will address the most popular of these models, the public cloud. In the following sections we will discuss the other three deployment models. Small and medium sized businesses typically use the public cloud as their primary computing source. Large corporations use the public cloud as an addendum to their in-house computing resources. In a public cloud model the organization could choose any one of the three types of services discussed earlier.

One of the benefits of cloud computing is in its ability to provide both computing and storage services as the need arises. In a classic example, Jeff Bezos of Amazon highlighted the rapid growth of Animoto company in requiring computing power that grew from 50 servers to 3500 servers over a three day period (Bezos 2008). Many companies need the ability to have on demand storage as the amount of data gathered is increasing rapidly. According to an IDC study the projected data growth by 2020 will be in the order of 40,000 Exabytes. Putting this number in the more commonly understood unit of gigabytes, it would be the equivalent of 40 trillion gigabytes (IDC 2012). Given this explosion of data and the need for large corporations to store large volumes of data, cloud computing is the natural solution. The same study by IDC points out that by 2020, the IT investment will drop from $ 2.00 to 0.20 per gigabyte. Given this reduced investment scenario by companies in IT infrastructure and the need to store very large volumes of data, a public cloud would be a natural choice since the cost of public cloud is relatively cheaper than a private cloud.

All of the benefits of cloud computing discussed in Section 3 apply to the public cloud deployment model. Small and medium sized businesses are able to subscribe to basic computing hardware and software in the public cloud. The pay-as-you-go model of cloud helps these businesses to concentrate on their core strengths and use the necessary cloud resources as the need arises and pay only for what they need. For large organizations there is the inevitable demand fluctuation and so investing in computer hardware to meet the peak demand is not the optimal solution. In such situations the public cloud comes handy for large organizations to use. Large organizations are able to use the public cloud for their non-sensitive data. One of the important things being considered by large organizations with regard to storage needs is the explosive growth of data. Data centers are key to the success of large organizations and the storage needs of data centers are met by using public cloud. The IDC Study estimates that in the case of large organizations data growth would be by a factor of six by 2010. Since then the technology has significantly improved in facilitating information sharing and rapid dissemination of geo location data through the use of social networks. To keep pace with this type of storage need will put a dent in corporate budgets. Associated with data storage is the need for ensuring security, privacy and availability of the stored data. Since the storage solutions offered by Amazon Web Services through their S3 (Simple Storage Service) cloud service addresses the security, privacy, backup and availability aspects, many large corporations use the S3 cloud service. The pricing model used by Amazon makes it attractive for all types of businesses to consider cloud storage as it costs less than 10 cents per gigabyte of storage. The 2011 Forrester Research Report points out that using cloud storage is 74 % cheaper than storing data in-house. (Reichman 2011) We will address the potential drawbacks of public clouds later in this section.

Public cloud offers several benefits. Foremost among these is the simplicity of use. Organizations that need computing services from a public cloud service provider such as Amazon Web Services is able to sign up for the service online easily. The provider offers both hardware and software resources that the customer needs on a pay-as-you-go model. The customer is able to benefit from the automated manner in which resources are allocated and deallocated. The service is available in a reliable manner on a 24×7 basis (Li 2011). The only requirement for the customer is the availability of a high speed internet connection. Many public cloud service providers carry compliance certifications such as HIPAA, SAS 70 and PCI. These certifications would help the customer provide the necessary documentation for their own compliance requirements.

One of the major concerns for many businesses is the public cloud security (KPMG 2013). It is a fact that public clouds host multiple tenants in one server using the virtualization concept. Customers feel that lack of control over the hardware, combined with the possibility of someone else accessing their data is a cause for concern. Since access to the public cloud is via an internet connection, customers are limited to the speed that they get through their ISP. Traditionally businesses will have the ability to depreciate their assets, including computer assets. However, use of cloud computing prevents the customer from depreciating any computer assets as they do not own the infrastructure. Sometimes customers feel that they are locked in to a cloud service because the cost of moving to a different provider is high due to lack of standards. At this time cloud service is still maturing and so many providers use proprietary technology to store data. Consequently, any customer migrating out of one provider to another provider will have difficulty in the usability of their data due to storage formats. This is a bigger problem for small businesses which tend to use SaaS more than PaaS and IaaS. Medium and large businesses that use PaaS and IaaS have a higher level of control over the storage aspects and so moving to a different cloud provider would not be that difficult. Overall, the benefits that the cloud offers far outweigh some of the disadvantages mentioned above.

2.6   Private Cloud

The concept of private cloud implies some form of ownership. It is true that the customer has higher level of control when using a private cloud. This is an expensive service compared to the public cloud. It would be affordable only to large businesses both from the infrastructure perspective and the system management perspective. There are four types of private clouds. In a typical *private cloud* the organization hosts the cloud in one of their data centers behind the corporate firewall. The architecture is akin to the use of intranet whereby the organization leverages the techniques of internet but limits the access to the content to internal employees only. In this sense the business uses the cloud technology but limits the users to its internal employees. The second type of private cloud is one that is managed by a third party provider. In this case the organization still owns the infrastructure in one of their data centers but management of the facility is with a third party. In this case the private cloud is called '*managed private cloud*' signifying that the infrastructure belongs to the organization but managed by someone else. In the third model, the '*hosted private cloud*,' a cloud service provider provides the infrastructure needed and manages that infrastructure. The benefit to the customer in this case is that scalability, demand elasticity and availability are guaranteed by the cloud service provider and because the servers are not shared with other organizations there is greater security. The fourth model is the '*Virtual Private Cloud* (VPC).' The VPC service is offered by a traditional cloud service provider in a multi-tenant environment (Microsoft 2012). In this section we will look at all these types of private clouds.

One of the major benefits of cloud computing is the economies of scale whereby the cloud service provider is able to leverage the servers through the use of virtualization. The benefit to the organization is the availability of large computing resources that they can use and pay for as needed. The organization stands to gain in the form of no investment in the infrastructure cost. Using the private cloud described above some of these benefits is lost because the organization has to invest in the infrastructure. Moreover, the infrastructure availability does not scale up quickly as in a true cloud. The use of the term 'cloud' in 'private cloud' is justified by the use of cloud concepts such as virtualization. Thus, the organization using the private cloud first overprovisions the infrastructure with the anticipated demand in mind and sets it behind the corporate firewall. Access is limited to the employees of the organization and much automation is provided in the form of enabling virtual servers on their physical servers. Since the organization controls both the infrastructure and access to the systems it is able to better meet the compliance requirements and have a higher level of security than in a public cloud. In this scenario the organization is able to manage the private cloud because it has a large IT workforce that could be dedicated to managing the private cloud. All the expenses of managing an information system would be applicable along with the cost of the infrastructure that requires overprovisioning. Examples of such private cloud users would be large organizations such as IBM, Cisco and Verizon.

In the second model, the 'managed private cloud,' a third party provides the management of the cloud service on internally owned infrastructure. The managed cloud service provider uses virtualization technology, thus eliminating the risk of multi-tenancy. There is a general myth that private cloud would be more costly compared to the public cloud. However, multiple studies have shown that it is not the case. For example, in a 2011 study, James Colgan points out that the cost of public and private cloud are almost the same, with the private cloud coming slightly below that of public cloud in cost. Here the comparison involves a fully private cloud and a leased private cloud with a public cloud (Colgan 2011). Colgan's analysis shows that assuming a 50% server utilization rate the cost for a private cloud is comparable to a public cloud. Likewise, in an EMC study in 2013, Chuck Hollis reports that in a typical usage, private cloud cost would be approximately $ 28,000 per month while for the public cloud with a single availability zone the cost would be approximately $ 33,400 per month (Hollis 2013). The same study also points out that if the usage pattern is more storage intensive then private cloud is cheaper than public cloud. On the other hand, if the usage pattern is more compute intensive then public cloud is cheaper than private cloud. Two of the prominent third party providers in managed private cloud are Hostway and Peer1 Hosting. Both these companies offer VMware based solutions for management.

In the third model, the 'hosted private cloud,' the cloud service provider provides dedicated servers to the organization, thus eliminating the concerns of multi-tenancy. This type of service offers all the benefits of cloud computing such as scalability, demand elasticity, availability and security. Hosted private cloud is also referred to as Leased private cloud. In this model the organization pays a slightly higher cost for the use of dedicated servers. Since the cloud service provider could not expect

to allocate these dedicated servers for other purposes during idle times, the customer's utilization rate would be considered 100 % of the server time, irrespective
of any lower percentage use. In this context we compare the cost details provided
by Colgan and find out that the public cloud would be significantly cheaper than
the private cloud because the infrastructure for the public cloud would be available
for other uses whereas the private cloud infrastructure would not be usable for other
purposes during any idle times. Some of the hosted private cloud providers are
Amazon EC2 Dedicated, IBM SmartCloud Enterprise, and Rackspace Cloud.

In the fourth model, the 'Virtual Private Cloud,' the cloud service provider provides virtual servers on multi-tenant hardware with VPN (Virtual Private Network)
access to their customers. Thus, the customer has certain added level of security in
their access to their virtual servers. This service is much less expensive compared
to the other three types of private clouds but more expensive than the public cloud.
Subscribers choosing VPC should use the IaaS type of cloud service since in that
case they would control all settings in their virtual servers. Since access to their
virtual servers is via VPN they get higher level of security than in the public cloud.
In this case the third party would be a bonafide cloud service provider with all the
advantages of the cloud service such as scalability, demand elasticity, availability,
and storage. The organization accesses the cloud using a VPN connection thereby
assuring security. Some of the Virtual Private Cloud service providers are Amazon
VPC, VMware, Microsoft Private Cloud, IBM SmartCloud Enterprise Plus, and
Rackspace RackConnect.

In the private cloud, a large organization which has a data center to manage, is
able to use large amounts of storage and computing power dedicated to just their
organization only. The private cloud facilitates the large organization to handle demand elasticity for computing similar to the public cloud provider using server
virtualization. However, in the case of storage the private cloud provider needs to
invest just for that purpose. This is not a major hurdle since organizations always
have their data storage needs grow and any capital investment in storage would be
worth the expense for the growing needs of the organization.

It is worth noting that the EMC Corporation, a multinational company specializing in data storage, predicts that by 2016 nearly 96 % of all Enterprise applications would either run on a private cloud or a Virtual Private Cloud. Likewise, of all
general applications nearly 61 % of them would run on a private cloud versus 39 %
that would run on a public cloud. Thus, the above cost observations concerning the
private cloud are important considerations for a business to consider in selecting a
cloud service provider.

2.7 Hybrid Cloud

Hybrid cloud is a natural evolution from the two service models discussed in the
preceding two sections. The hybrid model uses both proprietary computing resources that the organization manages directly and the public cloud for some of the

computing requirements, especially the ones with varying demands on resources (Bhattacharjee 2009). The typical hybrid cloud uses the Infrastructure as a Service type to build the public cloud part of the hybrid cloud. Since the organization controls the use of the infrastructure on the cloud they have total freedom to move applications between on premise and off premise. By having the public cloud component in the architecture the hybrid cloud offers the cloud advantages of scalability, availability, demand elasticity and pay-as-you-go model. Hybrid cloud is suitable for large businesses or niche businesses with a compute intensive system that would experience demand fluctuation (Piff 2012). Major cloud providers such as Amazon, VMware and HP all provide hybrid cloud service.

Most organizations today need the benefits of a hybrid cloud as new applications emerge and they need a way to test their viability before incorporating them with their systems and processes. For this reason the hybrid cloud management should be an extension of the private cloud management so that the organization could move applications seamlessly between the two parts of the hybrid cloud. Typically organizations are wedded to controlling their resources. With hybrid cloud technology using the Infrastructure as a Service model is able to offer the control the organizations desire. As discussed in the previous section many organizations prefer the use of private cloud for security and control considerations. The cost analysis discussed earlier shows that the private clouds are actually cheaper than the public cloud. Since the private cloud users are mostly large businesses the cost benefit advantage would not translate to small businesses.

One of the key benefits of hybrid cloud is the ability of the organization to keep on premise the sensitive applications and move to the public cloud the other applications. Using this blended approach the organization is willing to give up some control on its IT resources. It is worth noting that IT organizations tend to develop all the resources that they need. However, in the case of cloud technology the IT departments are bypassed by business units in acquiring the third party resources that they need in order to meet their goals. This reality has introduced the ability to test third party applications thereby opening up the opportunity to embrace new methods. In this way one could say that hybrid cloud helps future-proof the organizational IT by providing an avenue to introduce new techniques. This blend of private and public clouds clearly opens up the possibilities for application developers to come up with new methods that could be tested on a public cloud and eventually get adopted by private clouds.

Finding acceptance of hybrid clouds requires having the tools for the different development groups, IT personnel, and Quality Assurance team members to communicate well. One such tool is Eucalyptus, the open source software. The appeal of Eucalyptus is its compatibility with Amazon Web Services (AWS). This is a significant plus since much of the public cloud offering is dominated by AWS and a hybrid cloud needs the ability to interact well with AWS standards. The other open source software available for hybrid cloud is HyperTable. The HyperTable concept was developed by Google. The primary goal of HyperTable is to implement a scalable and distributed storage system that would deliver high performance. The main application envisioned for HyperTable is database processing. Already the database approach is migrating from standard SQL to NoSQL. Traditional database lookups involve keeping a hash table without the data itself being ordered on a key. In the HyperTable approach the data is stored in order based on a particular field such as URL. HyperTable uses a multi-dimensional table of information for rapid lookup. The HyperTable is implemented using a compiled code thereby providing greater performance speed. In a way this concept is similar to the MapReduce concept developed by Google whereby parallel processing is enabled using multiple servers simultaneously. With the cloud technology providing servers on demand organizations that adopt the hybrid cloud approach will be able to leverage the cloud to use the HyperTable. Being an open source software the organizations need not outlay significant funding for HyperTable (HyperTable 2013).

In a hybrid cloud environment of internal IT and use of public cloud computing causes some management control problems. The cloud itself consists of integrators, aggregators, third-parties, and vendors. Often businesses face a vendor lock-in for services because there are no universal standards governing the cloud technology. This concern is significantly alleviated by the use of IaaS service model. New services and niche players constantly emerge in cloud computing but lack of portability from one cloud provider or vendor to another makes it difficult to take advantage of future opportunities. The pace of change in cloud computing is difficult to predict. It is becoming increasingly evident that hybrid cloud is businesses could rely on because it offers the necessary control for part of their systems held internally and benefit from the cloud technology for non-sensitive applications. Another major concern of hybrid technology deals with jurisdictional constraints. Conflicting controls and legal requirements are a constant challenge since the cloud provider in one jurisdiction may not be aware of the legal requirements in another country or region. Also, the security controls of the vendor may not be in sync with the internal requirements of the business.

2.8 Community Cloud

The community cloud concept evolved when businesses in a particular sector such as automotive, energy, financial and health care realized that they need specialized applications that are not applicable to other sectors. The community cloud offers the benefits of public cloud computing but restricted to a particular industry segment and the security features of a hosted private cloud. Since the members of the community cloud are leading companies in that industry there is the mutual respect and protection for the services offered through the cloud. For example, a community cloud for the health care sector could focus on HIPAA compliance and the associated need for patient data protection and privacy. A traditional email service such a Gmail or Yahoo Mail will not be able to provide the differential security needed by certain types of email. For example, a communication between two doctors on

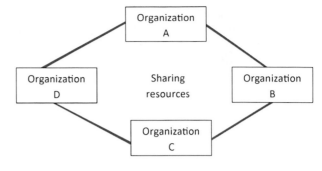

Fig. 2.3 Federated community cloud

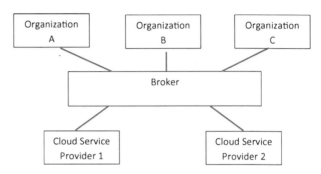

Fig. 2.4 Brokered community cloud

a community cloud would not be part of the general email community and as such will be able to provide greater protection for the content. People might be already aware that some financial companies make their confidential communications available only through their network to which the customer must login. However, the financial company will be able to alert the customer of the available communication through the public email system such as Gmail or Yahoo Mail. There are two basic types of Community Cloud models—Federated and Brokered. The underlying assumption is that businesses in the same sector only participate in this type of cloud. In the Federated model companies belonging to the same sector participate in the community cloud whereby any unused computing resource in one organization is used by another member organization on demand. Figure 2.3 illustrates a Federated Community Cloud. In the Brokered model a trusted third party serves as the broker and interfaces with the various community cloud members. The broker is responsible for procuring the various services needed by the industry sector and makes them available to all the members. Figure 2.4 illustrates a Brokered Community Cloud.

The community cloud is a closed system, available only to member organizations. The specific sectors identified earlier are large enough to manage a community cloud. An in depth analysis of the community cloud concept shows that it evolved from the techniques developed by vertical market segments, especially the automotive sector. The major benefit to the organizations belonging to the community cloud is that they will have greater cost savings in using the applications needed in that sector. The Brokered model supports this feature better than the Federated model.

The Federated community cloud model has the benefit of sharing the computing resources of member organizations when they are idle. However, the liability issues with regard to such processing are not settled, especially when there is a service outage in the middle of a processing. Another issue that is also open is to the responsibility of the member organization about keeping up the system and how the time share will be paid for. An example of a Federated Community Cloud is the joint project between Mt. Sinai Hospital of Toronto and the Canadian Government to make available a fetal ultrasound application to 14 area hospitals. This example highlights the benefits of a community cloud because hospitals generate 15 TB of data annually and this data growth is projected to grow at 31 times by 2020. When hospitals have such large volumes of data it becomes cost prohibitive to store all that data internally. Instead, using a community cloud the hospitals are able to not only save on storage costs but also share data with other hospitals.

The Brokered model is much better in handling the liability issues in the community cloud. The Broker is responsible for contract settlements with the providers, provide the necessary trust for the members for use of the system, and resolve disputes. Broker will also be able to provide the necessary data for the members for meeting their compliance obligations. In the case of cloud services one concern is how the cloud service provider will be able to mitigate the risks. The members need not be concerned about this aspect in a community cloud with a Broker present because the Broker will be able to take care of the risk mitigation. Moreover, the Broker will be able to provide several additional value added services. For example, the financial services industry requires low latency communication because they deal with buy and sell orders for stocks traded in a stock exchange. There are several Brokers who serve the community cloud segment of cloud computing. Some of them are Optum Health Cloud by United Health Care for the health care sector, IGT Cloud by International Gaming Technology for the gaming industry, and Global Financial Services Cloud by CFN Services for the financial sector.

Cloud computing as a whole is maturing rapidly and there are many service providers available globally. It gives vendors an opportunity to distinguish themselves by focusing their services in the community cloud for a specific sector. This availability of a large number of vendors in the community cloud segments by industry provides an opportunity for the industry members to take advantage of the community cloud offerings with the added security features not present in a general cloud offering. In this connection it is worth noting a report by the major international consulting firm McKinsey in which they identify opportunities and constraints based on the number of service providers in that vertical segment. The report's main theses are captured in Fig. 2.5 below (Stuckey and White 1993).

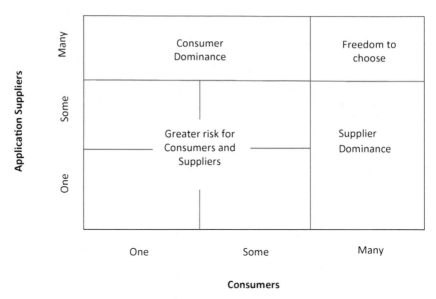

Fig. 2.5 Supplier—consumer options in community cloud

2.9   Storage Services

Earlier in this chapter we considered the three basic types of cloud services—SaaS, PaaS and IaaS. Analyzing the cloud computing usage landscape it is easy to note that cloud is heavily used for storage by many companies. The cloud storage serves two purposes for many companies. First, companies need to backup data and it is economical to use cloud for this purpose as the data is stored in encrypted form thereby providing the company the necessary security protection. Since the data backup is needed for a specific period of time companies could sign up for certain amount of storage such as 1 TB and reuse the space as needed. In this case there is no need to access the stored data on a regular basis and so the expectations for this data are purely for retention for a specific period of time. This is the most used type of storage service on the cloud, especially for disaster recovery, off site data protection and archiving. Second, companies could use the cloud for real time storage, also known as standard storage, as well. In this case companies have to look at the read/write speed of the cloud provider. If the company requires low latency in accessing the data then it would be more costly than medium and standard latency. The standard latency would be 10 ms for a gigabyte of data access. Since cloud storage offers the same type of benefit with regard to the pay-as-you-go model, companies may choose to follow the pay-as-you-go model. However, since the amount of storage needed is predictable, companies may opt to sign up for a certain amount of storage for a specific period of time such as one year at a lower cost than the pay-as-you-go model cost. Typically, Amazon and Google charge less than 10 ¢ per gigabyte of storage per month for standard storage. If the storage is used for backup only with no frequent access needed, then the cost comes down to about 7 ¢ per gigabyte.

Amazon Web Services has a specialized storage service called Simple Storage Service (S3) that is very popular. Typical latency rate for onsite data is 5 ms. So, companies use onsite storage for data that is needed often and with low latency. Other types of data that could tolerate medium to high latency such as 25 to 100 ms are moved to the cloud. The reason for higher latency rates in the cloud is two-fold. First, data is accessed over the internet and so there are intermediate routers and switches that have certain inherent delays built-in. Second, companies have a standard bandwidth access such as 100 MB over their corporate network to the cloud. For these reasons the 100 ms latency rate is not considered high. To improve performance cloud service providers use the deduplication concept. Deduplication means that only one copy of a data is stored and all other applications that require the same data point to the storage area for this data. Deduplication is extremely use-ful in email applications when an attached file is distributed to multiple users via email. Instead of storing the attachment in every Inbox, the deduplication allows one copy of the file to be stored in a common storage area and all the Inboxes that need this file simply point to this common storage area. Thus, storage service in the cloud not only provides basic storage but also provides efficient storage. Besides Amazon's S3 storage service, two other important storage services are Microsoft Windows Azure Blog and ATT Synaptic Compute as a Service.

Multiple studies have shown that corporate data storage in the cloud exceeds 60%. Hybrid approach to data storage is becoming popular whereby large compa-nies store most frequently accessed data onsite and the other data in the cloud. The data growth in companies is significant and with branch offices of a large company storing data onsite makes it more expensive and inefficient. For this reason many companies are switching to hybrid storage in the cloud. Gartner Research projects that by 2018, 80% of all corporate data will be on the cloud, with the rest being used for mission critical applications onsite. Gartner Research further emphasizes that it is not easy to compare the cloud storage costs with onsite storage because there are several other factors that go into the calculation of Total Cost of Ownership for the onsite storage (Gartner Research 2013).

2.10   Summary

In this chapter we have looked in detail the three cloud computing types—SaaS, PaaS, IaaS. In this analysis we pointed out the benefits and drawbacks of these services and their popularity and how they support security in the cloud. Also, we considered the four cloud deployment models—Public, Private, Hybrid and Com-munity. Through this analysis we were able to identify the strengths of each service model and give examples to illustrate that all these models are popular. The cost discussion on public cloud versus private cloud clearly showed that contrary to people's perceptions, the private cloud is more affordable and cost effective. However, one should note that the private cloud option is available only to large companies because of the cost outlay needed in order to consider private cloud. We concluded the chapter with a look at the cloud storage. It is clear from this discussion on storage that many companies use cloud for data backup, archiving or disaster recovery. It is estimated that a large percentage of corporate storage would move to the cloud by 2018 based on the access needs, data availability and cost considerations.

2.11   Review Questions

1. Explain what is meant by cloud computing and why it is attractive to businesses.
2. Explain SaaS, its benefits and drawbacks. Give examples of major service providers offering this service.
3. Explain PaaS, its benefits and drawbacks. Give examples of major service providers offering this service.
4. Explain IaaS, its benefits and drawbacks. Give examples of major service providers offering this service.
5. Explain Public cloud, its benefits and drawbacks. Give examples of major service providers offering this service.
6. Explain Private cloud, its benefits and drawbacks. Give examples of major service providers offering this service.
7. Explain Hybrid cloud, its benefits and drawbacks. Give examples of major service providers offering this service.
8. Explain Community cloud, its benefits and drawbacks. Give examples of major service providers offering this service.
9. Explain how Storage as a Service helps businesses use the cloud for data backup and recovery.

References

Bezos, J. (2008). Animoto company computing resources need. http://animoto.com/blog/news/company-news/amazon-com-ceo-jeff-bezos-on-animoto/. Accessed 8 Nov 2013.

Bhattacharjee, R. (2009). An analysis of the cloud computing platforms. MIT MS Thesis, Cambridge.

Colgan, J. (2011). Public vs private cloud cost comparison. http://www.xuropa.com/blog/2011/03/21/public-vs-cloud-costs/. Accessed 10 Dec 2013.

Combs, K. (2012). What is infrastructure as a service? Microsoft Technet Blogs.

Fielding, R. (2000). Architectural styles and the design of network-based software architectures. University of California—Irvine Doctoral Thesis.

Gartner Magic Quadrant. (2013). Magic quadrant for cloud IaaS. Research Report.

Gartner Research. (2013). How to calculate the total cost of cloud storage. Research Report.

Hollis, C. (2013). When public cloud is not cheaper. http://chucksblog.emc.com/chucks_blog/2013/03/when-public-cloud-isnt-cheaper.html#sthash.8K4RwMmV.dpuf. Accessed 9 Dec 2013.

Hurwitz, J., Kaufman, M., & Halper, F. (2012). *Cloud for dummies*. New York: Wiley.

HyperTable. (2013). http://www.hypertable.org. Accessed 11 Dec 2013.

IBM. (2012). Exploring the frontiers of cloud computing. Whitepaper.

IDC. (2012). The Digital Universe in 2020. http://www.emc.com/collateral/analyst-reports/idc-the-digital-universe-in-2020.pdf. Accessed 8 Nov 2013.

KPMG. (2013). Cloud takes shape—Global cloud survey. http://www.kpmg.com/Global/en/IssuesAndInsights/ArticlesPublications/cloud-service-providers-survey/Documents/the-cloud-takes-shapev3.pdf. Accessed 8 Dec 2013.

Li, A. (2011). Comparing public cloud providers. *Internet Computing, 15*(2), 50–53.

Loudhouse Research. (2010). Survey of IT decision makers. http://www.mimecast.com/Documents/Surveys/barometersurvey.pdf. Accessed 8 Nov 2013.

Microsoft. (2012). Microsoft private cloud: A comparative look at functionality, benefits, and economics. White Paper.

Piff, S. (2012). The age of cloud will be Hybrid, IDC. Whitepaper.

Reichman, A. (2011). File storage costs less in the cloud than in-house. http://media.amazonwebservices.com/Forrester_File_Storage_Costs_Less_In_The_Cloud.pdf. Accessed 8 Dec 2013.

Srinivasan, S. (2014). Is security realistic in cloud computing? *Journal of International Technology and Information Management, 23*, 1–2.

Stuckey, J., & White, D. (1993). When and when not to vertically integrate. McKinsey Consulting Report.

Chapter 3
Understanding Cloud Computing

Abstract Cloud computing has become very popular because it offers a wide range of computing services to a large group of businesses and individuals. Businesses have many choices to select a cloud service provider. The process of selecting and deploying the cloud service requires careful thinking on the part of the business to weigh the benefits against the possible drawbacks such as lack of control over the computing resources, applications and data storage. In this chapter we discuss in detail the many benefits and drawbacks of this service. Small and medium sized businesses tend to benefit from cloud computing because they are able to use sophisticated computing services without a large cash outlay. Individuals benefit from the ability to store and share information using the cloud. It is important for businesses to evaluate the various cloud service providers against their offerings and their fit with the business. The analysis in this chapter highlights the things that a business should look for in contracting with a cloud service provider. Even though cloud services are known for their high reliability and availability, over the past 5 years there have been several well-publicized outages. Because of the outages several other niche service providers who depended on other major cloud service providers were unable to deliver their service. We conclude this chapter with a detailed analysis of several cloud outages over the past 5 years that have eroded the confidence of businesses on the availability aspects of cloud computing as a 24×7 service. This discussion highlights the importance of service availability to build trust.

Keywords Cloud computing · Benefits · Drawbacks · Outages · Trust · BYOD

3.1 Introduction

Cloud computing as a technology is growing rapidly worldwide. Numerous small and medium sized businesses that could not afford to have an attractive web presence are now able to offer sophisticated services to the customers over the internet thanks to cloud computing. As discussed in Chapter 1, cloud computing technology came into being only in 2006. In less than a decade this technology has attracted the attention of the common consumer. This technology is benefiting individuals in sharing their photos and videos with family and friends. Companies like Apple, Dropbox, Flickr, Google, Microsoft and others offer free cloud storage for indi-

S. Srinivasan, *Cloud Computing Basics,* SpringerBriefs in Electrical and Computer Engineering, DOI 10.1007/978-1-4614-7699-3_3,
© Springer Science+Business Media New York 2014

viduals to share content with their friends and family. A recent report by Strategy Analytics shows that music and video dominate the usage by individuals in posting the content on the cloud for sharing with others (Strategy Analytics 2013). Apple is leading the way in music content with an overall share of 27 % of the users. Other leading providers of cloud storage are Dropbox (17 %), Amazon (15 %) and Google (10 %). From an individual user's perspective, Google's YouTube service is the leader in video content sharing on the web. This type of service alone is used by millions of individuals worldwide, making cloud computing an easy to use global technology.

The primary benefit of cloud computing for small and medium sized businesses is financial in nature. Investments necessary to have a reliable IT service has kept many prospective entrepreneurs from creating online ventures. On the web, businesses large and small look alike. Cloud computing is providing entrepreneurs the opportunity to try their ideas out, with IT services no longer holding them back as a barrier to entry. The major beneficiaries of cloud computing are small and medium sized businesses as this new concept provides them an opportunity to try out high-end services with no up-front cost, allowing them to use the pay-as-you-go model. Large businesses also use cloud services a lot, but because they have the necessary financial and people resources their use of cloud service is at a high end, which we will discuss in detail later in this chapter.

Looking at the growth of cloud computing over the past decade, it is easy to notice that this service is popular because the users need not acquire any new infrastructure. The basic requirement is the availability of access to the internet, which many people around the globe have today. It is worth noting that people tend to use a technology if it is simple to use. For example, the popularity of browsers is attributed to their availability on all computers stored today. Likewise, on the security side, many internet users support eCommerce because the SSL encryption is native to the browsers and the customer need not do anything special to acquire this software application. Moreover, in many developed countries the availability of mobile devices has skyrocketed and the telecommunications providers are able to use technologies such as LTE (Long Term Evolution) to offer high speed data service over mobile devices. This feature has made the cloud service even more useful to the customers. These are some of the driving forces behind the popularity of cloud computing.

The leaders among cloud service providers are Amazon, Google, Rackspace, and Microsoft. Even though all these providers are US based, cloud computing is a global technology and there are numerous niche cloud service providers in every region of the world. We discussed in detail the three major types of cloud services in Chap. 2. These are Software as a Service, Platform as a Service and Infrastructure as a Service. Software as a Service is used largely by small and medium sized companies. This accounts for nearly 60 % of all cloud services. For many SaaS is synonymous with cloud computing. With BYOD (Bring Your Own Device) becoming popular among businesses, many small and medium sized businesses are trying to take advantage of the cloud using BYOD. Platform as a Service and Infrastructure as a Service each have a market share of around 18 % each. The rest of the cloud market place is shared by other niche providers focusing on a specific aspect of cloud such as security. Major cloud service providers target consumers more than enterprises. Cloud service providers such as Microsoft, Google and Dropbox offer free services for storage up to a certain level. Invariably the customers will need more storage and so they are highly likely to sign up for paid storage. In some cases higher level of availability is guaranteed for premium services that people sign up for. Small and Medium sized businesses need specialized services related to data backup and recovery as well as integration of several of the services that they use over the cloud. Customers pay for these services on a recurring basis, providing a revenue stream for cloud service providers. This is one way free services bleed into paid services for cloud service providers.

Large enterprises stand to benefit more from cloud computing, although of a different nature. Large enterprises manage data centers and the IT paradigm shift referred to earlier mean more in the context of accessing data from the data centers. In this context private clouds have been introduced where the benefits of storage management and elasticity in demand for computing services are the key drivers. Moreover, the cloud technology also offers high level of reliability and availability of systems without significant capital layout. Often, the benefits of cloud computing are realized by taking a hybrid approach. The hybrid approach gives the large organizations the ability to manage their IT centers and at the same time expand their computing capacity without large capital investment by utilizing the cloud resources. This is especially useful to meet seasonal peak demands with hybrid clouds. Organizations with seasonal high demands that could benefit from hybrid clouds are in the entertainment industry around holidays, sports networks with on-demand service and tax service providers.

3.2 Advantages of Cloud Computing

Cloud computing has many benefits to offer. In this chapter we will look at the details of these benefits and how they benefit businesses of all types. Businesses plan to use cloud computing primarily because they could *transfer all the management issues* associated with managing an information system. From the perspective of small and medium sized businesses this boosts the ability of the business to focus on their core strengths and enjoy all the benefits of having an information system available on demand. Closely related to this is the matter of having the computer system available for business use all the time. This is known as service *availability*. This is measured in system uptime, which in today's terms means 24×7 availability. Cloud service providers tout their system uptime using a simple metric called 'number of 9s.' This means that customers expect the cloud computing system to be available almost non-stop. The three 9s availability means that the service provider guarantees that the computer system would be up 99.9% of the time. This is a significantly large uptime. When a company owns its own Information System it will have to make available significant resources in order to have the three 9s uptime. This is a costly phenomenon. It is important to realize that when a company owns its own Information System it has to bring the system down every week for routine maintenance. But when the service is moved to the cloud, as a user the expectation for system uptime increases. From the details presented in Chapter 2, the 99.9% uptime translates to a total downtime of nearly 1½ min per day. Cloud companies strive for at least four 9s, i.e., 99.99% system up time for their cloud service. The four 9s uptime translates to a maximum downtime of at most 8 seconds per day. In order to guarantee such a high uptime the cloud service providers have to invest heavily in redundancies and automation in service allocation. When an individual organization's computing system is down it only affects that one organization and its customers, which usually is a smaller number. However, when a cloud service provider system is down, since they serve numerous customers, it has a much larger impact. As we will discuss later in this chapter on outages, the unknown for the customer is the total downtime, as the cloud service provider is unable to provide this estimate during an outage.

Another major benefit of cloud computing is its ability to provide limitless server capacity. This is also known as service elasticity or *demand elasticity*. Cloud companies architect their system using the concept of virtualization to make limitless servers available. Virtualization means a single physical device is partitioned into multiple virtual machines (VMs) and made available to the customer. The main physical device is called the 'host' and each VM is called a 'guest.' Each guest requires minimum 2 GB storage. Each VM can be a virtual server. If the host has at least a Quad core processor with at least 1 TB hard disk space, then ten virtual servers can be created on the single host, each with 100 GB storage allocated. For today's workloads the 100 GB storage allocation is small. In reality the number of virtual servers per host depends on the type of workload planned per server. Typically the hosts come with at least 32 TB hard disk capacity and four multicore processors. The goal of this illustration is to point out that a single host could provide multiple servers since the typical server utilization is only around 20% of its capacity. Even with all servers functioning at 50% of their utilization capacity the load demand will not be taxing the host's processors. This illustration is intended to show that using the server virtualization scenario any cloud customer is given the appearance that the number of servers that they have access to is limitless. This point is well illustrated by the experience of Animoto company that experienced tremendous service growth that it required to go up from 50 servers to 3500 servers over a 3-day period (Bezos 2008). This was made possible by Amazon Web Services for Animoto. Businesses consider having limitless computing power availability an important aspect in opting to use cloud computing versus managing their own computing system. We mentioned earlier that this benefit is called demand elasticity. This means that a business could use a certain amount of computing resource at one time and release part of that at other times when the demand for such a resource is less. This ability to ramp up or down in the need for computing resources is what makes cloud computing attractive to many businesses (European Union Report 2009). This is known as *scalability* and businesses benefit significantly from this feature. An associated benefit here is the sustainability of the computing resource being used.

Cost is an important metric for all businesses. Cloud computing supports the pay-for-what-you-use model. This is also known as *pay-as-you-go model*. Given the need for demand elasticity discussed above, the pay-as-you-go model enables the customer to not invest in expensive computing hardware that they use rarely.

From the cloud service provider perspective they are able to pass on the cost benefits to the customer because of economies of scale. This benefit of cloud computing makes it a utility for the customers. The cloud service provider has the advantage of choosing to locate their servers in an area of the country where the cost of electricity is lot less. For example, the per kilo watt hour rate in northwest USA is about 8 cents whereas the rate in northeast USA is about 18 cents (Energy Information Administration 2013). Because of the advancements in communications technology location of cloud servers does not pose any service constraints since all customers access their cloud infrastructure via the internet.

Cloud service architecture is *distributed* for availability and reliability. Because of this feature cloud services are able to provide *24 × 7 customer support*. From the customer perspective the services offered are location agnostic. Further, the customer is able to *use any device* to access the cloud service. For example, businesses that manage their own information system know that they have to manage a different set of APIs in order to make the information available on mobile devices and PDAs. Since cloud service providers cater to the needs of a large and varied set of customers they make the necessary infrastructure investments to make their services accessible through all types of devices. From the business perspective this is a major plus because of the prevalence of BYOD sentiment among many employees. Thus cloud computing supports the increased *mobility* of the workforce. As seen from the Animoto example above, cloud services significantly *reduce infrastructure deployment time*.

Businesses that use the cloud service know that they have the ability to change their service mix rapidly. For example, a customer needing to test one of their applications on a different platform such as Mac or Linux will be able to get their resource quickly and test their application. This type of *agility* is supported by cloud service providers and valued very much by the customers. We discussed in the last chapter three types of cloud services – SaaS, PaaS, IaaS. The PaaS service enables the customer to choose any platform that they need to either offer or test their services. Thus, cloud service supports *multi-platform availability*. The availability of all types of resources on the cloud enables the customer to reduce their capital expenditure significantly and shift it to operating expenditure. Since all aspects of management of the information system for the business is shifted to the cloud it frees up internal resources in the organization.

Cloud services are used extensively by businesses for storage. Some of the associated actions needed with storing data is backup and security. Businesses should protect the stored data by using encryption. The encryption key should be kept internally by the business. One common problem with cloud storage is the proprietary format used by the cloud service provider in the absence of global standards. Because of this many businesses are locked-in with the provider. Instead, businesses could contract with a third party specializing in storage whereby the customer preferred application data will be preserved in that application format. This helps during retrieval of such data in the form of speed. Cloud services also help with third party integration of some of their services. For example, a company using a third party for payroll service would find it easy to integrate their internal data for payroll with the payroll service since they will be providing similar service to several other businesses as well over the cloud. This cloud aspect would be a cost saver for businesses.

Businesses tend to look at the various components of their business from the Total Cost of Ownership (TCO) perspective. For example, if they were to host their own computing system then it requires not only infrastructure but also experienced people to manage the system. Thus, the cost of managing a computing system extends beyond the infrastructure costs and so using cloud service significantly lowers the TCO. Moreover, businesses are able to have access to more advanced web services such as online chat, online credit card processing and integrated web site for the business. When the demand for more cloud based services arise it becomes easy to acquire the new resources and implement them quickly. Since much of this is done under user control, implementation is quick and easy. For example, a business requiring additional computing power and storage space during the tax season is able to acquire these services over the web and release them when the demand wanes. The best feature of cloud in this regard is the high degree of automation used where by the service requests are fulfilled immediately by the cloud service provider. Businesses pay for the services that they use. However, if the demand for a particular type of service such as storage is needed for a specific period of time such as a year, then the cloud service offers subscription based pricing for the extended duration. The use of cloud also helps the cloud service provider to offer additional specialized services required by several small businesses through their Software as a Service. Cloud becomes a focal point for several small and medium sized businesses to find additional opportunities for collaboration with other service providers. For example, a business focused on providing custom furniture needs a way to get their product to the customers and in this case they can collaborate with a transportation company using the cloud service to be a partner.

Overall, it is important to realize that the use of cloud service has numerous benefits to offer the businesses, the foremost being service availability on a 24×7 basis, unlimited resources, meeting demand elasticity, pay for what is used, access to high-end services and sustainability. Because of server virtualization, businesses are able to request on demand many servers and have them deployed immediately. Besides the infrastructure provided, the cloud service also provides several applications such as Microsoft Office 365, Google Apps and the Customer Relations Management software by Salesforce.

3.3   Drawbacks of Cloud Computing

Cloud computing is widely accepted as a new approach to using information technology in a cost effective way. In the previous section we have highlighted the many benefits of cloud computing. In this section we will discuss in detail the drawbacks of cloud computing. The purpose of mentioning these drawbacks is for the customer to be aware of the issues associated with cloud computing. These will help the customer choosing to use cloud computing to incorporate the factors that might be problematic for them in the contract.

Multiple surveys of cloud users have consistently shown that *cloud security* is the top concern for all cloud customers (Intel 2012). This concern stems from multiple layers. We will look at the security concern from these various aspects that the customer focuses. The primary concern for many users is the *loss of control* over the hardware and data (Hashizume et al. 2013, Onwublio 2010, Zissis and Lekkas 2012). By design the cloud infrastructure is to be accessed using the internet. Customers who are accustomed to having the computing infrastructure in-house feel that since they are not aware of where the hardware is they have a feeling that it may be insecure (Mather et al. 2009). This is a false assumption because the cloud service providers, given their size and technology deployed, have adequate resources to provide physical security for their hardware and provide additional electrical power protection (Cloud Security Alliance 2011).

The second concern is about the *location of data storage* in the cloud. Cloud service provider is responsible for storage and backup of data. The service providers deploy enough redundancy to guarantee a high degree of service availability. In order to achieve this, the service provider has to choose locations that are geographically apart. This inherent design feature makes it difficult for customers to know where their data is stored, not just they are available when they need it. Some countries require that businesses must store customer data within the country. This is because of differing legal requirements in countries with respect to privacy protection. European countries and Canada have explicit laws that require data to be stored within the country. The primary reason cited for this is the concern that any data stored in US could be within the reach of US government because of PATRIOT Act (Ars Technica 2011).

The cloud security concern extends to state governments as well. When data is stored in a central location there is greater risk of the data being stolen. When such a data loss occurs it is massive in scale. Because of this some states such as Massachusetts and Nevada have passed laws that require protection of data pertaining to their residents when stored in the cloud. Other states enacting similar laws are California, Virginia, Iowa and South Carolina. These laws require that the cloud service providers take adequate precautions to protect the data. Even though there is no federal law requiring data protection for data stored in the cloud, the state laws described above pertain to residents of those states. These laws would apply to all businesses that provide service to residents of these states. Even if the business is not based in one of these states they must still comply with the data protection and as such the reach of these laws essentially make them applicable to the entire nation. One precaution such a business could use involves encryption for stored data. Customers would also benefit by choosing an encryption of their own before their data is stored in the cloud. In this scenario the customer should be able to safeguard the encryption key in their internal system. Since SaaS applications on the cloud handle data in the clear, any business using an encryption scheme to store data must present the data in clear for use by the application. Srinivasan (2014a) discusses these and other details in his paper on security for cloud computing.

One of the reasons for customer concern about cloud security is *multi-tenancy* of customers on cloud servers. Cloud service providers offer virtual servers to customers by providing computing power using servers that host multiple clients. The concern is that because data belonging to different customers reside on the same physical server there is a possibility that such data could be accessed intentionally or accidentally by other customers. Data that persists on systems after their use is known as data remanence. Potential hackers could subscribe to a large amount of storage space on a virtual server with the idea of accessing data that remains in the storage area after use by another customer. This violates the confidentiality aspect of an information system for clients. This is exacerbated by the popularity of Software as a Service (SaaS) on the cloud. With SaaS, multiple clients using the same third party software could discern more information from accidental access to other clients' data on the same physical server. Furthermore, cloud service providers handle data in clear in their SaaS offerings. Earlier we considered encryption as a possibility but small and medium sized businesses that use SaaS cloud service will not have the financial resources to handle encryption.

Customers who choose cloud service opt for this service because of the need to have service available at all times. To a very great extent cloud service providers are able to provide 24×7 service *availability* but at times the service experiences outage that extends beyond the acceptable limit for the service uptime guarantee promised. We review some of the major outages in cloud service later in this chapter. Another drawback is the *cost* savings to customers by using cloud service. Most of the customers, especially small and medium sized businesses, use SaaS cloud service. Without the capital investment expenses for IT systems, cloud customers benefit initially. Studies have shown that this benefit extends for the first 2 years of a service contract for SaaS. Beyond that period customers pay more for cloud service than what it would cost them to host the service internally (Ward n.d.). The primary reason for this is that the business cannot depreciate the cost of IT infrastructure since it does not own them. Gartner Research reports that SaaS services are not purely for use when needed and rapid deployment alone. Most SaaS applications require a predetermined level of commitment from the customer and certain specialized applications would take a significant time to become available for use on the cloud. These realities should be taken into account when selecting cloud service.

Cloud service's popularity is based on its *availability over the internet*. Cloud customers need not acquire any specialized hardware or software in order to use the cloud service. This simplicity of access over the internet adds to the security concerns because the internet is not designed with security in mind. Cloud customers can overcome this drawback by using the Virtual Private Network (VPN) service offered by the telecommunications provider. However, this adds to the cost of cloud service and so small and medium sized businesses will not be able to afford the VPN service.

Two other aspects that are also considered as drawbacks in the use of cloud computing relate to trust and compliance. *Trust* requires the provider to have well defined policies that follow known standards. For example, the cloud service provider should be able to provide the customer with third party certifications as to their service controls in the form of SAS 70 Type II Audit statement. Additional certifications to the ability of the service provider to protect the hardware and software for HIPAA and SOX compliance would help build trust. The service provider should

Table 3.1 Summary of Cloud Service Concerns

Concern	Cause	Remedy
Security	Traditional security methods do not apply. Access over public internet.	Develop issue-specific protection and VPN access
Availability	Service outages	Identify causality and address solutions such as power supply backup
Loss of control	Inherent in design from customer perspective	Choose PaaS or IaaS service to address specific areas such as application or system access
Trust	Lack of standard measures to make best practices visible to customers through third party validation	Apply third party validation and acquisition of standard certifications such as SAS 70
Cost	SaaS cost benefits drop after 2 years	Choose applications carefully
Multi-tenancy	Applications and data of multiple customers reside on the same physical server because of virtualization	Use encryption to protect data
Data storage location	Cloud service provider needs redundancy and so multiple copies are kept in geographically dispersed locations. Customer can choose primary storage location from the various data centers of the provider.	Keep sensitive data in-house and store only other data in the cloud
Compliance	Cloud service operations may not be transparent to the customer	Periodically collect log data and access data from provider to verify access control. Cloud provider should obtain certifications such as SAS 70 and HIPAA.

be able to provide the log data for the customer in an automated manner so that the customer can meet their obligation for *compliance* requirements.

We conclude this section with Table 3.1 in which we summarize the major concerns about the cloud service discussed above. The level of the concern varies with the size of the business. Large businesses which have additional resources to manage their won computing centers are much more concerned about using a cloud service than small and medium sized businesses. Small and medium sized businesses are more concerned with the cost of service than sharing the service on the same server with other businesses.

3.4   Essential Components for a Cloud Contract

Cloud services are here to stay. More and more businesses, especially small and medium-sized businesses, depend on cloud computing heavily. In this section we will highlight the important factors that a cloud customer should consider before deciding on a cloud service provider. Over the past 8 years numerous cloud service providers have emerged, with the goal of providing several niche services. The industry is handicapped by the absence of global standards for cloud computing even though the service is truly global in nature. It is important to recognize that a provider who claims to be a cloud service provider is indeed one such. This is important because there are many service providers who focus on storage or file sharing only and do not own any infrastructure of their own. They lease their infrastructure from major service providers such as Amazon Web Services and offer a niche service. Thus, a customer must first verify whether a service provider has their own infrastructure in offering the cloud service or depend on another service provider. It is perfectly legitimate to use such a service if it would meet the business needs. There are precedents for this type of service in the energy and telecommunications sectors.

It is important to note who the major cloud service providers are and what they have to offer. Amazon Web Services (AWS), Google App Engine, Microsoft, Rackspace, Apple, SalesForce, Dropbox, and IBM are important cloud service providers. Of these major service providers, AWS has the largest market share by providing cloud service under SaaS, PaaS and IaaS types as well as the deployment models of public cloud, private cloud and hybrid cloud. The two major services of AWS are Elastic Compute Cloud (EC2) and Simple Storage Service (S3). Google App Engine provides PaaS and IaaS services besides the popular Gmail service. GoogleDocs is an application service that provides standard office products such as word-processing, spreadsheet and presentation software. GoogleDrive provides a popular file sharing service. It is also integrated with GoogleDocs so that documents could be edited in GoogleDocs format. Microsoft through its Windows Azure platform provides PaaS service. Its Office 365 web service integrates basic office products like Word, Excel and PowerPoint as a SaaS service. Microsoft's SkyDrive provides cloud storage and integrates with its basic Office Suite for editing documents. Rackspace is a global cloud service provider that focuses on superior customer service for all its cloud offerings. The other major cloud providers focus in one or more of these areas. For example, Apple's focus is on music using iCloud. SalesForce focuses on Customer Relationship Management (CRM) application only, with an associated social networking called Chatter. Dropbox has been a pioneer in cloud storage for file sharing. Over the last year it has grown exponentially with an additional 100 million users. Given the vast infrastructure that AWS has, Dropbox uses AWS S3 service to store the actual files of its 200 million users. Dropbox's nearly 10,000 servers are used to store the metadata of its users for the billions of files being stored. IBM focuses on providing managed cloud service.

By its very design cloud service works over the internet and so any customer with access to the internet will be able to sign up for cloud service. The speed of the internet connection often decides the response time with the cloud service because the cloud service providers have the ability to match the communication speeds of most Internet Service Providers (ISPs). In Table 3.2 we summarize the main features to look for in such a service.

A potential customer considering cloud computing as an option either to supplement their internal computer information system or to use as their primary information system should consider the essential features described above. Whether the

Table 3.2 Features to look for in cloud service

Feature	Requirements
Availability	Service availability on a 24 × 7 basis is important. This is measured by a unit such as three 9s (for 99.9 % availability) or four 9s (99.99 % availability). In order to provide such high availability the provider has to maintain extensive infrastructure with redundancies.
Security	Internal controls in place to support security best practices. Data pertaining to third party audit of system for protecting hardware and software.
Ease of use	Well tested interfaces that are connected to service provisioning in an automated manner. Customer should be able to handle demand elasticity through allocation and deallocation.
Feature Set	Customer should be aware of all the different types of services such as on demand computing, support for basic cloud types and different deployment models, storage and device independence for access.
Mobile Access	All services offered through the cloud should be accessible through the different types of mobile devices
Support	Technical support through multiple channels such as web, phone, chat, email
Compliance	Support for customer's need to acquire necessary system data such as login data for all users, system uptime information, system patch updates, any security vulnerabilities

business is small, medium or large, once using cloud computing, its availability all the time is very critical. For this reason the contract should spell out clearly the financial remedy in the event of a failure to uphold the system availability. As mentioned in Chapter 1, a guarantee of system availability at four 9s only allows for a total downtime of 52 min per year, including any routine maintenance times. On the security side, the service provider should be able to identify all their internal controls for system management, all privileged users with access to customer data and the level of multi-tenancy planned for the servers. In order to build trust with customers the cloud service provider should obtain SAS 70 Type II Audit certification and the like to show that their internal controls work and they are compliant with system management policies. The customer should be able to try out the system prior to signing the contract to test the ease of use of the system. Among the services that a customer might need would be SaaS applications. If this is the primary goal of the customer then they should find out what predetermined levels of service guarantee are expected from the customer. As mentioned earlier, the cloud computing service is a pay-as-you-go model but yet the cloud service provider expects the customer to use a minimum level of access and service. This is important for the customer to know because it involves cost commitment. If SaaS service is planned the customer must be aware that the SaaS service is cost effective for them only for 2 years. So they should plan accordingly in the contract as to the continuing service cost beyond the 2 year period.

Customer needs vary widely when it comes to applications. The potential customer must carefully evaluate the available applications to see if they would meet their needs. If additional applications are needed then the customer must be able to get an estimated time duration by which the application would be available on the cloud. As discussed earlier it might take a significant amount of time for the cloud service provider to make available any specialized software. For many businesses application availability on mobile devices is important. Even if the business may not need it for their employees, businesses should make their information accessible on mobile devices. This requires the cloud service provider to make the necessary changes to the content in order to make them available on mobile devices. Cloud computing introduces many new types of risks and so the cloud customer will be faced with the need for technical support. For this reason the contract should include appropriate response time in the Service Level Agreement (SLA). Finally, customers may have the need to comply with certain laws such as the Health Insurance Portability and Accountability Act (HIPAA) or the Sarbanes Oxley Act (SOX). In order to meet the compliance requirements the customer might need log data pertaining to system access and the system management information. Such data should be easily available for customers in an automated manner. Before deciding to use a particular cloud service the customer should consider all these aspects and decide if the cloud service will be able to meet their needs.

3.5   Major Outages

Cloud service provider's computing system is a very large system with multiple redundancies. Over the past five years several of the large cloud service providers have experienced service disruption or in layman's terms service outage. In this section we will explore the details on these service outages and the important lessons learned both for the service providers and the customers.

Amazon Web Services (AWS) is the largest cloud service provider in the world with popular offerings such as Elastic Compute Cloud (EC2) and Simple Storage Service (S3). On April 21, 2011, AWS experienced significant service latency in its Virginia data center. AWS has eight data centers around the world for storage redundancy. With three data centers each in US and Asia Pacific region and one each in Europe and Latin America, AWS provides EC2 and S3 service worldwide. Each region consists of multiple Availability Zones. Each Availability Zone consists of multiple Elastic Block Store (EBS) volumes. The Virginia data center, during a routine system maintenance and upgrade, encountered a system management error due to erroneous routing. This human error caused the system to not respond for service requests from major businesses. The outage lasted nearly 4 days. Amazon provided 10-day service credit to all affected businesses (Amazon 2011). This outage showed the vulnerability of many businesses that depend on a single cloud service provider. In 2012, the Virginia data center was affected again in June and October by utility power failures due to thunderstorms and failure of backup generators. The Amazon outages affected large businesses such as Netflix, Instagram, Pinterest, Foursquare, Quora and Reddit.

Gmail is Google's popular email service that is used by over 425 million users globally. Gmail is a free service but is also part of Google's premier GoogleApps service. When an outage occurs in Gmail, even if it is for a very short time, it has significant impact on users because of the number of users involved. Gmail experienced several outages in 2008 that lasted as much as 30 h. Microsoft suffered a major outage in its Sidekick application in Fall 2009. This outage, which lasted nearly 1 week, Microsoft lost all data as well from the Sidekick application that helped users access email, calendar information and personal data through a cell phone. This service was provided by T-Mobile and it issued credits of up to 3 month data charges. The Microsoft email service, Hotmail, also suffered an outage at the end of 2010 that lasted up to 3 days. In this case the outage was caused by programming error. These email outages show that the errors are caused by humans.

Another global cloud service provider is SalesForce.com. This company provides internet-based CRM service to thousands of companies worldwide. The company's service experienced an outage in January 2009 that lasted less than an hour but impacted over 177 million transactions worldwide. Likewise Terremark, a global colocation and cloud hosting provider with its 13 datacenters worldwide, experienced an outage in its vCloud Express service in March 2010 for 7 h and a more recent outage in October 2013 in its data center that affected the Healthcare. gov web site for several hours. Another significant cloud outage was with the GoDaddy.com web hosting service. GoDaddy hosts 5 million web sites and manages 50 million domain names. On Sept. 10, 2012 it had an outage that lasted 6 h and was caused by corrupted data in router tables in its internal systems.

Rackspace is a global cloud service provider with a special focus on customer service called Fanatical Support. Rackspace had multiple outages in 2009, some resulting due to power problems. The outages in its largest data center in Dallas occurred in June, July and November 2009. In December 2009 the problem was due to a routing loop problem between its Dallas and Chicago data centers. In January 2013 Rackspace had a Distributed Denial of Service (DDoS) attack that affected its email servers for a brief period of time, causing communication outage with its customers.

Dropbox revolutionized file sharing using cloud storage in 2007 when it was founded. Since then the company has grown to more than 200 million users worldwide and numerous businesses globally that subscribe to its Dropbox Business service. Dropbox users store one billion files daily and access their files using 500 million mobile devices. Dropbox uses nearly 10,000 servers to store file metadata and store the actual files on AWS S3 servers. In 2013 Dropbox suffered multiple outages, some due to problems with AWS S3. In some instances the outages lasted over 15 h. One disturbing fact is that Dropbox is not willing to share the cause for the outages so that customers could take alternative steps to protect their content availability (Dropbox 2013).

Cloud customers, in spite of much frustration over outages, continue to depend on cloud computing. They believe that with good communication from the cloud service providers they will be able to plan for actions during outages. The major cloud service providers decided to improve communication with the customers through the use of dashboards that provided real time updates on their cloud services. We describe next some of these dashboards and the information that they convey. Amazon Web Services has created a dedicated web page called Service Health Dashboard with system update on all its cloud services. The data for this page comes from automated data feeds from the various cloud services within Amazon. This Dashboard features data by Service Centers in order for customers to know about the status of the storage location that they had selected. This is a direct result of Amazon to build trust with their customers by providing with automated system status information following several outages (Amazon 2013). Google, which also had multiple outages over the past 5 years, has created an Apps Status Dashboard web page with real time information about the status of all its Apps products, including Gmail. Its visual is not as useful as other Dashboards from Amazon and Rackspace because it is all blank when there are no problems reported (Google 2013).

Microsoft has taken a slightly different approach in providing the status information for its cloud services. In a brief Dashboard called Service Status, Microsoft provides the status information on all its cloud services without providing any additional detail (Microsoft 2013). Then it has more detailed system information on each of its major products such as Windows Azure, Office 365, etc. that are accessible by the user account and specific to a data center. From the general user perspective this information should be publicly available to all without having a Microsoft cloud service account (Microsoft 2013). Rackspace maintains a real time System Status web page with automated data feed from their servers. This information is not as detailed as Amazon's but certainly provides trust building information to the customer. This is in response to several outages over the past 5 years that resulted in loss of some customer trust due to lack of real time system information during an outage (Rackspace 2013).

We summarize the major cloud outages over the past 5 years and their known causes. This information is presented in Table 3.3.

3.6   Trust Enhancers for Cloud Service

Trust is a critical aspect of cloud service. Software applications running on the cloud expect the users to trust their applications for proper handling of customer data. The problem arises here when an application, due to lack of strong authentication requirements, introduces additional risks while handling sensitive data. The most popular type of cloud service is SaaS. The way SaaS applications handle data is in the clear. Consequently the cloud service providers have to develop methods to protect their infrastructure from external attacks. Conveying this information to the customers in a transparent way helps build trust in their service. Building trust is a time consuming process. In this section we will review several recommendations that would enhance customer trust of cloud services.

An effective way to build trust is to develop transparent policies that help the customers manage their core areas of strength. In this regard the cloud customer needs data pertaining to customer access of any information that they have on the web. By making this information available in a proactive way the cloud service earns the trust of cloud customer. Another important area of concern for customers

Table 3.3 Summary of cloud outages. (Source: security, trust and regulatory aspects of cloud computing in business environments)

Date of outage	Cloud service provider	Outage duration	Cause of problem
Aug. 11, 2008	Gmail	5 h	
Oct. 16, 2008	Gmail	30 h	
Mar. 13, 2009	Windows Azure	22 h	
Jun. 29, 2009	Rackspace	45 min	Power interruption
May 11, 2010	Amazon	1 h	Power interruption due to external event
Apr. 21, 2011	Amazon	3 days	Automatic backup configuration
Feb. 28–29, 2012	Windows Azure	1 day	Leap year processing
Jun. 14, 2012	Amazon	6 h	Power interruption
Jun. 28, 2012	Salesforce	5 h	Storage tier issues
Jul. 10, 2012	Salesforce	12 h	Power interruption
Jul. 26, 2012	Windows Azure	2½ h	
Sep. 10, 2012	GoDaddy	6 h	Corrupted data in router tables
Oct. 22, 2012	Amazon	6 h	Fix for failed hardware cascaded
Oct. 26, 2012	Google Apps	4 h	
Nov. 8, 2012	Microsoft	8 h	Maintenance issues and network failures
Nov. 15, 2012	Microsoft	5 h	
Jan. 10, 2013	Dropbox	16 h	
Jan. 28, 2013	Facebook	2 h	Denial of Service
Feb. 1–2, 2013	Microsoft Office 365	2 h	Error in routine maintenance update
Feb. 22, 2013	Windows Azure	12 h	
Mar. 14, 2013	Microsoft	16 h	
Mar. 18–20, 2013	Google Drive	17 h	Network software problem
Apr. 23, 2013	Apple iCloud	Several hours	
May 30, 2013	Dropbox	1½ h	
Sep. 13, 2013	Amazon	2 h	

is security. The cloud service provider not only should protect customer data but should also communicate the controls that they have in place to protect the customer data and processes. The use of encryption for stored data helps alleviate some of the security concerns for customers.

Cloud service providers should strive to have in place mature processes and internal controls that meet the industry standards in health care and finance. Validating the processes and controls should be done through a third party that can provide SAS 70 Type II Audit certification. Additional certifications that the cloud service provider could seek are in HIPAA, SOX and GLBA. If the service provider is able to have these certifications then it becomes easy for the cloud customer to trust the cloud service more. Also, it would help in the customer meeting their compliance requirements.

One core aspect of cloud computing is the control of infrastructure by the cloud service provider. In order to enhance customer trust the cloud service provider should be able to document the uninterrupted chain of service provider control over the application interfaces to the infrastructure. Cloud service providers use the virtualization technology extensively resulting in multi-tenancy both for computing power and storage. Tools such as Hy Trust that combine virtualization technology with protection of cloud infrastructure facilitate customer trust. Thus the goal of cloud service provider should be to incorporate trust building tools in their cloud service. Another important area for building trust is the availability of system as per the service agreement. Even though the cloud service providers take extraordinary precautions to provide uninterrupted service, the fact of the matter is that all major cloud service providers have encountered service outages for extended periods of time that violate the service agreement. Providing cost credit to the customer helps with trust but the focus should still be in having uninterrupted service.

We conclude this section with a set of recommendations for best practices that enhance customer trust (Srinivasan 2014b). These are:

1. Building customer trust requires making relevant log data available to customers on demand and in an automated manner
2. Identify the list of employees at the cloud service provider with privileged access to networks
3. The cloud service provider obtains industry recognized compliance certifications from regulators and government
4. Cloud service provider should make available audit data of their systems
5. When an incident such as cloud outage or a denial of service attack happens the provider should have well established disaster recovery procedures
6. Cloud service provider's uptime claim should be independently verifiable
7. Cloud customer should have the ability to choose their data storage locations from among the service provider data centers
8. Access to newer encryption technologies provide a higher level of security for customer data before they leave their control

3.7   Summary

The primary focus of this chapter has been in bringing forth the basic building blocks of cloud computing. After a review of these building blocks we discussed the benefits cloud computing has to offer. This discussion included well known examples of cloud services such as email that are benefiting a large number of customers worldwide. In order to present a balanced view of cloud computing we discussed extensively the many concerns with regard to cloud computing. Considering together the benefits and drawbacks of cloud computing it is clear that the benefits far outweigh the drawbacks. In this context it is worth noting that when health care services are offered in a rural setting, the beneficiaries are more interested in receiving the health care for a better quality of life than getting concerned about the possible privacy violations of having the health care information on the cloud. We followed up with some of the major outages that occurred over the past 5 years in cloud services. A closer look at these outages clearly showed that cloud services are as vulnerable as other services because there is a human element to it and errors are bound to creep. In some of these outages the service providers were affected by natural disasters that interrupted the power supply for a large part of a region. We pointed out how many of the service providers have learned the importance of communication with their customers during an outage. This has resulted in dashboard data being made available over the web concerning the availability or otherwise of the various cloud services offered by the provider. Finally, we discussed the importance of sharing information with the customers in order to build trust. Trust building takes time and the service providers must keep their customers informed of the best practices that they are implementing.

3.8   Review Questions

1. What are the common cloud services that individuals and businesses use, especially as a free service? Explain how these free cloud services bleed into paid services.
2. What are the major benefits of cloud computing and how are they perceived by businesses of various sizes?
3. What are the major drawbacks of cloud computing and how are they perceived by businesses of various sizes?
4. What should businesses look for in a cloud contract and how the cloud service providers are unwilling to modify their standard contract?
5. Describe the impact of four major cloud outages that affected businesses.
6. How can cloud service providers enhance the trust placed in their services by customers?
7. Explain five different best practices that would enhance customer trust of a cloud service?

References

Amazon. (2011). Summary of the Amazon EC2 and Amazon RDS Service Disruption in the US East Region. http://aws.amazon.com/message/65648/. Accessed 22 Dec 2013.

Amazon. (2013). Amazon Service Health Dashboard. http://status.aws.amazon.com/. Accessed 23 Dec 2013.

Ars Technica. (2011). PATRIOT Act and privacy laws take a bite out of US cloud business. http://arstechnica.com/tech-policy/2011/12/patriot-act-and-privacy-laws-take-a-bite-out-of-us-cloud-business/. Accessed 21 Dec 2013.

Bezos, J. (2008). Animoto Company Computing Resources Need. http://animoto.com/blog/news/company-news/amazon-com-ceo-jeff-bezos-on-animoto/. Accessed 8 Dec 2013.

Cloud Security Alliance. (2011). Security guidance for critical areas of focus in cloud computing v3.0. https://downloads.cloudsecurityalliance.org/initiatives/guidance/csaguide.v3.0.pdf. Accessed 22 Dec 2013.

Dropbox. (2013). http://www.dropbox.com. Accessed 23 Dec 2013.

Energy Information Administration. (2013). http://www.eia.gov. Accessed 20 Dec 2013.

European Union Report. (2009). Cloud computing: Benefits, risks and recommendations for information security. http://www.enisa.europa.eu/activities/risk-management/files/deliverables/cloud-computing-risk-assessment. Accessed 5 Feb 2014.

Google. (2013). Apps status dashboard. http://www.google.com/appsstatus#hl=en&v=status&ts=1387805946273. Accessed 23 Dec 2013.

Hashizume, K., Rosado, D.G., Fernandez-Medina, E., & Fernandez, E.B. (2013). An analysis of security issues with cloud computing. *Journal of Internet Services and Applications, 4*(5), 1–13

Intel. (2012). What is holding back the cloud? http://www.intel.com/content/www/us/en/cloud-computing/whats-holding-back-the-cloud-peer-research-report.html. Accessed 20 Dec 2013.

Mather, T., Kumaraswamy, S., & Latif, S. (2009). *Cloud security and privacy*. Sebastopol: O'Reilly.

Microsoft. (2013). Online system status. https://status.live.com/. Accessed 23 Dec 2013.

Onwublio, C. (2010). Security Issues to cloud computing. In N. Antonopoulos & S. Gillam (Eds.), *Cloud computing: Principles, systems and applications*. London: Springer-Verlag.

Rackspace. (2013). System status. https://status.rackspace.com/. Accessed 23 Dec 2013.

Srinivasan, S. (2014a). Is security realistic in cloud computing? *Journal of International Technology and Information Management, 23*(1–2).

Srinivasan, S. (2014b). Risk management in the cloud and cloud outages. In S. Srinivasan (Ed.), *Security, trust and regulatory aspects of cloud computing in business environments*. Hershey: IGI Global.

Strategy Analytics. (2013). Cloud storage. http://www.informationweek.com/cloud/infrastructure-as-a-service/apple-dropbox-lead-cloud-storage-market/d/d-id/1109199? Accessed 15 Dec 2013.

Ward, S. (n.d.). 5 Disadvantages of cloud computing. http://sbinfocanada.about.com/od/itmanagement/a/Cloud-Computing-Disadvantages.htm. Accessed 22 Dec 2013.

Zissis, D., & Lekkas, D. (2012). Addressing cloud computing security issues. *Future Generation Computer Systems, 28*(3), 583–592.

Chapter 4
Cloud Computing Providers

Abstract Cloud computing is popular today because of the reliable services provided by major companies such as Amazon, Google, Microsoft, Rackspace and Terremark. In order to provide reliable cloud computing service the provider must invest large sums in infrastructure. Their architecture includes several redundancies. Each of these global service providers have several data centers spread all over the world. These data centers help cloud consumers meet their governmental requirements that cloud data must reside within the country or in the region. Moreover, these distributed data centers facilitate storage redundancy and help with low latency response. Also, these providers offer some or all of the three types of cloud services—SaaS, PaaS and IaaS. Besides these cloud service providers (CSPs), there are also several niche cloud service providers such as Salesforce, Apple, VMware, Dropbox and SoftLayer. These companies focus on specific services such as Customer Relations Management (CRM), music distribution, virtualization, storage and bare metal servers. Another set of businesses focus on providing third party coordination of service for businesses needing cloud service. In this chapter we describe the services offered by the major cloud service providers, niche cloud service providers and highlight the important role that these third party facilitators provide and how small and medium sized businesses could benefit.

Keywords Large CSPs · Niche CSPs · Third party facilitators · Service differentiation · Service focus · Service availability

4.1 Introduction

Cloud computing is a global phenomenon today. Most of the major cloud service providers are U.S. based while some of the niche cloud service providers are located in Europe, Australia and Asia. These cloud service providers have distributed data centers all over the world. The primary reasons for placing these data centers in various parts of the world are to meet some governmental requirements about their citizen's data not leaving their region or providing low latency service globally. In order to provide low latency, i.e., faster response time, the cloud data should be closer to where the cloud customers are located. The distributed data centers also serve the purpose of geographic separation for storage backup. Moreover, some of the U.S. laws like the USA PATRIOT Act and the disclosures about the NSA surveillance programs have resulted in many governments requiring that their citizen's data be located outside of U.S. Such requirements apply to Canada and countries in the European Union. Europe has Safe Harbor Agreement with U.S. which gives confidentiality protection for data stored in U.S. Besides the U.S.-Europe Safe Harbor Agreement, Switzerland has enacted additional privacy requirements, especially given their banking regulations (U.S.-EU Safe Harbor 2000; U.S.-Swiss Safe Harbor 2008). Thus, a major concern for many cloud customers is where their data is stored and how it is protected from governmental intrusion.

Cloud service providers have the ability to provide computing service at an affordable cost. For customers the cost savings alone is not enough. The service providers have to earn the trust of the customers. That is why knowing where the data will be stored and how the data will be handled are important for cloud customers. At this time there are no accepted global standards for storage or security aspects of cloud service. Because of that some customers are hesitant to use the cloud for fear of provider lock-in. Cloud service providers are trying to address these concerns by joining industry consortia like the Cloud Security Alliance (CSA), Cloud Industry Forum (CIF), and the Open Data Center Alliance (ODCA) (Cloud Security Alliance 2014, Cloud Industry Forum 2014, Open Data Center Alliance 2014). Both CSA and ODCA are U.S.-based non-profit organizations. CIF is based in Europe. These organizations are trying to develop standards that could be followed by all cloud service providers. Another effort in this regard is the Open Stack for cloud and Open Nebula. Open Stack is an open source software developer consortium that has developed a cloud operating system for massively scalable systems that support private and public clouds, two of the most popular cloud deployment models. At present only Rackspace and HP have adopted Open Stack and even in those adoptions there are variations and so there is not much interoperability option among various cloud services. OpenNebula is also an open source software for cloud management platforms that are scalable. As the cloud computing industry matures these efforts will pay off, enabling the cloud users to move freely between cloud service providers, just as there is number portability in telecommunications services.

Cloud computing is having a significant impact on ecommerce. Projections are that this trend will continue for some time to come (Gartner Research 2013). In order for the growth to continue in cloud computing adoption the customers should have the facility to switch service providers easily. Even though theoretically cloud computing promotes itself as a service that could be used on demand and customer pays for what they use, in reality customers will incur cost and time delay in porting their data between service providers. The reasons for this aspect to be looked at carefully prior to service initiation are many-fold. First, there is an added cost that the service provider charges to transfer data out of their storage. Second, the bandwidth available for transfer may not be adequate for a quick data transfer. Businesses depend on their data very much and the data storage grows rapidly for businesses. So, to reach a data storage size in terabytes is not uncommon. Third, many niche providers enter the cloud service provider market through the use of services such as Infrastructure as a Service and Platform as a Service from major cloud service providers. Given the low barrier to entry for being a cloud service provider there is a high likelihood of some cloud service providers going out of business is high. These factors should make the cloud customer to explore the time and cost aspects of switching service provider.

Given the rapid growth of cloud computing industry many small and medium sized businesses do not have the ability to assess the quality and reliability of provider assurances for service. To facilitate these businesses choosing the right type of cloud service for their needs and to provide additional value added services to customers, many Cloud Service Brokerages (CSBs) have emerged. These Brokerages are able to evaluate the many cloud service providers available and to negotiate contracts for common services such as email and data storage with service providers and pass on the savings to the smaller cloud customers. The value added service they provide is in the form of service integration among a variety of related services such as email and voice mail. The CSB service has grown into a multi-billion dollar service industry in itself. Many small and medium sized businesses value the advice and guidance of CSBs in selecting the right type of cloud service for their needs. In this chapter we will discuss in detail the various cloud service providers and their services. Also, we will look at several niche cloud service providers from U.S., Europe, Australia and Asia who have either a service, regional or language focus. We address the important role of third party cloud service facilitators, also known as Cloud Service Brokerages. We conclude the chapter with an analysis of the current state in the development of global cloud service standards.

4.2   Major Cloud Service Providers

Cloud computing's attractiveness to businesses involves the service availability and reliability. In order to provide both availability and reliability the cloud service provider has to make significant investments in infrastructure. Also, the service needs to have several redundancies in order to meet customer expectations. The unit to measure service availability is the number of 9s in its uptime guarantee. For example, major cloud service providers will be expected to have an uptime of three 9s meaning that their system should be available 99.9% of the time. This translates to a total downtime of nearly 9 h per year, including time for any routine maintenance. We use the three 9s criteria to classify cloud service providers as major cloud service providers. These major cloud service providers who are unable to meet the three 9s service availability provide service credit to the customers. In spite of the best efforts, cloud service providers have experienced several service outages during the past 5 years and so it would be unreasonable to expect a higher level of service uptime guarantee. If the service provider were to guarantee a four 9s service availability guarantee then that means that they will have to maintain their service all the time, except for 52 min in a year. The economic impact of cloud services both in commerce as well as in the cost of cloud services is significant. Gartner Research's projection is that between 2013 and 2016 businesses will spend $ 677 billion in cloud services. An IDC study projects that the revenue from public cloud services will account for $ 73 billion by 2015 (IDC 2013).

Based on the criteria set in the previous paragraph for major cloud services, the following companies would qualify to be classified as major cloud service providers based on either size or popularity of their offering with businesses:

- Amazon Web Services
- Google Apps
- Microsoft Windows Azure and Office 365
- HP Cloud
- Rackspace
- CSC Corp
- Verizon Terremark
- Dropbox
- Box

We discuss in detail the varieties of cloud services offered by these companies, the type of Service Level Agreement (SLA) these companies offer and their recent statistics on service availability over the past 2 years. The purpose of this analysis is to let the potential customer know the aspects of service that they should be aware of in choosing a cloud service provider.

Cloud service has been meeting the needs of organizations of all sizes for development and testing using multiple platforms. Given the growth of cloud computing over the past decade the businesses now have to come to depend on cloud services for routine work as well as in production systems. Managing a cloud service requires the provider to offer some or all of the following types of services: SaaS, PaaS and IaaS. These major service providers not only offer these types of services as well as support the four deployment models—public cloud, private cloud, hybrid cloud and community cloud—discussed in detail in Chapter 3. The focus of the major cloud service providers are in the public and private clouds primarily. The hybrid and community clouds are often the focus of niche service providers. In the following subsections we will discuss in detail the offerings of these major cloud service providers.

4.2.1   Amazon Web Services (AWS)

Amazon Web Services (AWS) is the largest of the cloud service providers. Its market capitalization is such the computing resources utilized by various businesses through Amazon are five times larger than all the other cloud services combined. These cloud service providers are all featured in Gartner's Magic Quadrant (Gartner Research 2013). AWS offers a free usage tier similar to the email services from Google, Yahoo and Microsoft. Businesses could use the free usage tier and test their needs and grow into other premium services offered by AWS such as Elastic Compute Cloud (EC2) and Simple Storage Service (S3). AWS provides a single point of contact for customers for all their cloud services. This is especially helpful when the customer is adjusting to the cloud environment.

Amazon launched AWS in 2006. It has steadily grown this infrastructure over the years. Amazon's overall investment in cloud infrastructure is approximately $ 12 billion, which is less than some of the other major cloud service providers. In spite of the lower investment AWS is able to offer their cloud services globally. It has several thousand customers spread over 190 countries. AWS specializes in having various regions strategically located around the world for data storage that the customer could choose. The existing regions are distributed as follows:

North America   3
Europe   1
Asia   3
Australia   1
Latin America   1

Moreover, AWS has a separate cloud service region for the US government. It is an isolated cloud both physically and logically. This cloud service is FIPS compliant and as such customers using this cloud for their service to the US government will be able to provide FIPS compliance data. To further facilitate customer choice in selecting the region for both storage and computing resources to meet their compliance requirements in certain cases AWS has created several Edge Locations within each region. For example, in the US East Region based in Northern Virginia there are 12 Edge locations in places like New York, Miami, Dallas and Atlanta that the customer could select.

AWS offers SaaS, PaaS and IaaS services as well as uses the public cloud and private cloud deployment models. AWS services are used by companies of all sizes. For example, Dropbox, which pioneered file storage and sharing in the cloud uses AWS's S3 service to store all the customer files, which run into several billion files. The demand elasticity that S3 service provides enables Dropbox to use as much space as needed to store all customer files, which come at the rate of 1 billion files per days. Dropbox uses its own servers as well but they are not used to store customer files. Besides Dropbox other notable large companies that use AWS cloud service are Netflix, Flickr and Pinterest.

Given the wide popularity of AWS, numerous customers worldwide depend on the availability of AWS across all time zones. So, even a small service outage at AWS has great impact on people's need for service globally. This became quite evident from several major outages at AWS over the past 5 years. Some of these outages were caused by power supply disruption to AWS while others were caused by human error. In every one of these instances AWS has credited the customers for loss of service because of the pay-as-you-go model it supports. AWS comes up with detailed post-outage analysis and makes its findings available for the customer.

AWS offers SLA with three 9s availability and a host of applications in its SaaS service. Its public cloud is very heavily used and it offers the US federal government exclusive and isolated cloud service. By locating their service regions around the world and providing customer the option to select a particular storage region it is facilitating customer compliance with respect to knowledge of where the data is stored. To facilitate customer's ability to meet their compliance requirements AWS carries multiple compliance certifications such as SAS 70 Type II Audit, FISMA, HIPAA and SOX compliance. AWS being the largest cloud service provider many software vendors have licensed their product to run on AWS's EC2 platform. Incorporating AWS' API has become critical for many third party providers who offer cloud management services.

4.2.2   Google Apps

Google Apps evolved over the past 5 years. The initial App was Google Docs which was launched in 2006. Traditionally people around the world use Microsoft Office products such as Word and Excel. Clearly many versions of these Office products are in circulation and at times people run into the difficulty of having version incompatibility in opening certain Office documents. Google Docs was designed to overcome this problem by providing document and spreadsheet creation and sharing capability on the web. The benefit of Google Apps was to provide the customer with an application that allows:

• Sharing files in various formats
• Uploading or downloading files in various formats
• Publishing files directly as HTML files
• Keeping the files in read-only format on demand

One of Google's original cloud service was its Gmail service. The Gmail service is used today globally by over 450 million users. Since then Google has expanded its Google Apps to include Gmail, Google Docs, Google Drive, Google Talk, Google Calendar, Google Video, Google Labs and Google Play. All these services are very popular with the public because they are free for the public. For example, the Google Drive comes with a generous 30 GB free storage capacity and Google Talk lets customers communicate over the web for free in spoken format. The latest App, Google Play, is an all-encompassing service that incorporates music, movies, video, books and news magazines. It is bound to have a significant following in the years to come.

Google's cloud infrastructure grew rapidly over the last 3 years. Today, Google's investment in cloud infrastructure is approximately $ 21 billion, far more than Amazon. Unlike Amazon, Google is into specific services such as Gmail and Google Talk. Moreover, much of Google's services are free and are supported by advertisements. Given this model Google has many competitors for specific services. For example, in music the largest competitor for Google Play is Apple's iTune, which is a pioneer in that field. In the case of online file storage, the competition for Google Drive comes from Microsoft's Sky Drive and Dropbox, both of which offer free service as well for the public.

Google had its share of cloud outages over the past 5 years. In some instances the outage has lasted as long as 30 h. Since millions of customers depend on Google Apps all over the world, even a small service disruption for 15 min in its email service will have significant impact on customers.

Google offers SLAs with three 9s availability. Google offers all three basic types of cloud services—SaaS, PaaS and IaaS. Its popularity as a cloud service provider is backed by its major services such as YouTube and Gmail.

4.2.3   Microsoft Windows Azure and Office 365

Microsoft offers cloud service across multiple types—SaaS, PaaS and IaaS. Windows Azure, launched in 2010, specializes in PaaS and IaaS services. Microsoft Office 365 specializes in SaaS services from the Microsoft Suite of products. Office 365, in combination with Microsoft's SkyDrive, enables business customers to easily share word documents, spreadsheets and PowerPoint slides among users in dispersed geographic locations. As a true cloud service with a global reach, Microsoft's cloud offerings provide the cloud benefits of scalability, demand elasticity and pay-as-you-go model for pricing. The SLA provided by Microsoft offers three 9s availability. Like other cloud service providers, Microsoft also experienced several outages over the past 3 years. Given the size of the organization some of the causes for outages show that their internal management controls are not adequate. For example, their systems did not plan adequately for leap year handling in 2012 and another failure was caused by simple failure to renew their security certificate. Since Azure supports database services using SQL as well as NoSQL, customers have the ability to run a variety of database systems over the cloud.

Windows Azure offers the customers a free usage tier which they could try first and then subscribe to other premium services offered by Azure. Because of the tight integration of the Azure platform with other Microsoft products such as Office 365, customers could use a PaaS or SaaS service using one of the Microsoft products. Windows Azure provides a single point of contact for their cloud services in order to enhance the product usefulness to the customer.

Microsoft entered the cloud market much later relative to other major service providers such as Amazon and Google. Yet, Microsoft's investment in cloud infrastructure is approximately $ 18 billion, which is more than AWS' in this regard. Since Microsoft is able to provide a host of commonly expected office productivity services such as email, Office Suite of products, it cuts down on adoption issues because most users are familiar with the products.

Microsoft provides dedicated Government cloud for various government agencies in US at the federal and state levels. Microsoft's government cloud service is FISMA, HIPAA, SOX, SAS 70, EU Safe Harbor Framework and ISO 27001 certified. Microsoft offers the federal government agencies dedicated public, private and hybrid cloud environments. Its data storage policy lets the agencies know where their data is stored. One of the benefits of the government cloud is that the government agencies are able to provide citizen services at a reduced cost. For example, NASA makes available large volumes of data it gathers from its space explorations for citizens to access through the government cloud. The state of Ohio provides real time traffic information to the motorists through the Buckeye traffic web site (Microsoft 2010).

Microsoft's service uptime guarantee is at three 9s like all the other service providers. Learning from the problems faced by Amazon Web Services in managing stored data, Microsoft has invested over $ 2 billion in new data centers around the world that are capable of running when electrical service is disrupted.

4.2.4 HP Cloud

HP offers cloud computing as a PaaS and IaaS service. The HP Cloud Compute is a public cloud. HP is one of two cloud companies that support OpenStack, an open source cloud operating system. The main purpose of supporting Open Stack standards is to support service portability. The HP's implementation of Open Stack is still evolving and there are incompatible components in the implementation of cloud service that makes interoperability with other cloud services difficult. The primary goal of Open Stack standards is to enable interoperability using third party providers. HP's Cloud Compute supports both Windows and Linux environments. Two of HPs notable cloud offerings are Converged Infrastructure and Cloud Maps. The HP Converged Infrastructure provides greater automation in service provisioning and supports multiple operating systems. The HP Cloud Maps integrates services from multiple vendors such as Oracle, SAP, VMware and SAS.

HP CloudSystems forms the basis for HP's cloud strategy. This is planned for launch in 2014, giving the users greater flexibility. The HP CloudSystems will enable the customers to burst into other public clouds offered by AWS, Windows Azure and niche service providers in France, Germany and Spain. The partnership with providers in European countries is to meet the customer expectations in those regions concerning data storage in those regions to meet their compliance requirements.

HP's SLA offers three 9s availability, similar to all the other major cloud service providers. As a new entrant into the cloud service market HP is trying to differentiate itself from others both in service requirements and pricing. AWS offers the same SLA as HP and in order to meet this SLA, AWS requires users to subscribe to two different Availability Zones so that the customer will not experience downtime. Moreover, AWS offers a 10 % service credit when their systems are not available. HP, on the other hand, requires only one Availability Zone subscription and service credits of up to 30 % when an SLA is not met. Some industry observers question HP's claim on SLA.

4.2.5 Rackspace

Rackspace is one of the largest cloud service providers in the world. Besides HP, Rackspace is the other company that supports OpenStack. It supports the Open-Stack operating system for the cloud but it is not fully interoperable with other cloud services. The goal of OpenStack is to provide full interoperability with all cloud service providers but it will take some more years for this technology to mature before all cloud service providers adopt the OpenStack operating system for the cloud. The Rackspace version of the OpenStack cloud is not compatible with other cloud services. The OpenStack service is gradually evolving. Rackspace is giving the customer the option of using the new Solid State Drive (SSD) technology which provides faster access to stored data. The SSD technology is significantly more expensive than the traditional Hard Disk Drive (HDD) technology. For cost comparison, it costs $ 0.075/GB in HDD whereas the cost is $ 1.00/GB in SSD. This makes the Rackspace cloud service cost higher than that of other cloud service providers for those that seek the faster speed for their mission critical services.

Rackspace is known for its customer support, known as Fanatical Support. Rackspace provides SaaS, PaaS and IaaS services. Rackspace provides an unusual guarantee of 100 % service uptime, except for routine maintenance. To provide such high reliability Rackspace partners with major vendors like Cisco and CA Technologies. It has a total of 10 data centers in 6 regions around the world. The redundancies built into the architecture include the services of 9 different network service providers with high bandwidth. Rackspace's investment in cloud infrastructure is approximately $ 2 billion, far less than other major cloud service providers such as AWS and Google. Through the use of network service providers such as CA Technologies Rackspace Support team is able to get full visibility of the customer network and could fix customer access issues quickly by interacting directly with the customer infrastructure rather than through service advisors.

Its database service uses container-based virtualization instead of the traditional server-based virtualization. This feature combined with high-capacity network and high speed SAN storage service provides customers with scalable high performance database systems. Using the open source MySQL software the customers get the cost benefit. One of the benefits of cloud service is the ability to eliminate redundancy in storage. This feature in the cloud is called de-duplication, which means that only one copy of a file is stored and all additional applications that need this stored file content simply have a pointer to the file storage. The de-duplication can be at the file level or block level. Typically when an email attachment is sent to several users the same file is stored by different users and de-duplication avoids this extra storage requirement. Rackspace uses both de-duplication and block level compression during data backup. It uses incremental backup beyond the initial backup which means that only changes to the files that occurred since the initial backup are stored. These features save significant costs for customers for data backup as the savings are 10–20 times when compared with the traditional storage technology.

Rackspace had service outages over the past 5 years due to power disruption. Its largest data center in Dallas, Texas, was affected multiple times over the past few years due to external events that caused the traditional electric utility service to be disrupted for a longer period of time and its backup system was not able to keep up with the power requirements. In case of service disruption due to outages, Rackspace offers 5 % service credit for every 30 min of downtime.

4.2.6   CSC Corp.

CSC Corporation is a large global cloud service provider with data centers in US, Europe, Latin America, Australia and Asia. Often customers need support services in order to manage their cloud deployments. CSC provides customer support in multiple languages such as French, German, Italian, Spanish and Mandarin. CSC's main cloud offerings are vCloud, BizCloud and CloudCompute. The use of vCloud from VMware shows CSC's close partnership with VMware. The BizCloud is an enterprise level private cloud meant for large businesses. BizCloud deployment for an enterprise through CSC requires 10 weeks to architect. This cloud based service is a physically segregated service offering high security and reliability. It saves the enterprise from committing funds for capital expenditure and instead uses the benefit of the cloud with pay-as-you-go model. Moreover, the enterprise could use the managed service option that CSC provides as a cloud service provider. CSC has standardized its offerings both in the public cloud and private cloud. Moreover, it uses the same pricing model in both of these deployment models. CSC offers SaaS, PaaS and IaaS services.

CSC's CloudCompute service is also an enterprise level service. It is offered as an Infrastructure as a Service giving the organization higher levels of control as to the types of applications that it wants to run on the cloud. This service could be run using any one of the 15 data centers distributed worldwide. Since CSC uses VMware's vCloud, all applications used in any of these data centers are fully portable. The CloudCompute service includes the basic cloud services for computing power, storage capacity and high speed networking capability. CSC's offerings are geared towards large enterprises and so small and medium sized enterprises will not benefit much from CSC's expensive services.

CSC offers Storage as a Service, similar to AWS' S3 service. CSC offers Government Cloud Service for US Federal government agencies in a secure, isolated environment. By offering both public cloud and private cloud services, CSC is able to meet customer needs for higher levels of security or lower costs (CSC 2014).

4.2.7   Verizon Terremark

Verizon as a company is a telco that has a large presence in the mobile communication market both for voice and data. Its Terremark division is focused on enterprise level cloud Infrastructure as a Service (Verizon 2013). These services are provided in North America, Latin America, Europe and Asia. Terremark partners with VMware in offering vCloud Express service which makes its service portable. Verizon Terremark's advantage compared to the other cloud service providers is in having the worldwide network connectivity provided by Verizon. Terremark capitalizes on this and offers colocation services to enterprises. Colocation enables the enterprises to host their mission critical applications on Terremark infrastructure that are located in data centers worldwide. The data centers are connected to multiple communication service providers. Terremark's public cloud service connects to AWS, Open-Stack and Rackspace's CloudStack. It also supports OpenStack's Swift standard. Terremark's cloud service supports over 450 different operating systems, including the major versions of Windows and various flavors of Linux.

Terremark provides government cloud service as well. This service is isolated from the rest of the public cloud offerings from Terremark and supports both NIST and FISMA standards. Terremark cloud service complies with multiple standards such as SAS 70 Type II Audit, PCI DSS and the European Union Safe Harbor requirements. Terremark also has the new SSAE 16 Type II Audit certification. SSAE (Statement on Standards for Attestation Engagements) came into force in 2011 and effectively replaces SAS 70 Standards and certified by the same organization. Terremark uses the SAN architecture for supporting high volume storage with rapid access needs. Its SLA includes three 9s availability (99.9%) for communication services through Verizon infrastructure and dedicated cloud services. For the Enterprise Cloud Service, Terremark provides SLA only at 99.5% whereas all the major providers such as AWS, Microsoft, Google and Rackspace provide 99.95% availability.

Terremark stands to gain from the availability of Verizon's communications infrastructure to provide reliable VPN tunnels for higher level security for cloud applications. Through its Infrastructure as a Service and high reliability, Terremark provides other businesses the necessary infrastructure to support their Platform as a Service. Even though Terremark focuses on large enterprises for its IaaS service, through the provisioning of infrastructure for other companies such as Engine Yard, Terremark provides access to medium sized businesses access to cloud services.

Security is a major concern for many businesses planning to use cloud service. Terremark's infrastructure physical security is well tested because of Verizon. Through the use of an enveloping technology called Cloud Switch, Terremark provides virtualized instances that are encompassed by the isolation technology. Thus, the entire cloud deployment of the customer is within a secure envelope that is encrypted end-to-end as the data moves through the Enterprise cloud. Thus the security feature in Terremark's service is easy to measure. It also supports integration of Big Data Analytics using the open source Apache Hadoop framework through its partnership with Cloudera. Since Big Data Analytics is becoming critical for many large enterprises, having this feature enhances the value of Terremark's cloud service.

4.2.8 Dropbox

Dropbox is a pioneering cloud storage provider with over 200 million customers and 4 million businesses worldwide. The only cloud service that Dropbox offers is storage. It manages over 10,000 servers to keep customer file metadata and uses AWS S3 service to store the actual customer files. Data indicates that customers upload 1 billion files daily. The large customer base is due to the availability of 2 GB free storage space, which can be upgraded to 100 GB for an annual cost of $ 99.

Dropbox provides 256-bit encryption for data in transit or rest. Since the actual files are stored with AWS, Dropbox benefits from AWS' high availability and reliability for customer file access.

Dropbox experienced several outages during the past 5 years for brief periods of time. It inconvenienced the customers a lot. Dropbox did not identify the causes for the outages, making it difficult for customers to place trust in the service. Dropbox has serious competition in the storage service from both Google via its Google Drive application and Microsoft via its SkyDrive application. Google offers 15 GB of free storage and Microsoft offers 7 GB of free storage compared to Dropbox's 2 GB free storage. Dropbox has simplified the storage and access process by developing an API that customers can download and install on their desktops. Beyond that the desktop folder serves as the link to Dropbox which makes it easy for the users to simply drag and drop their files in the desktop folder for cloud storage. Dropbox automatically syncs in the background the customer storage folders with their cloud storage. This ease of use has made this service more popular. Moreover, this service is available on all three major platforms—Windows, Mac and Linux.

Dropbox gives the customers the ability to access the stored files in any device, including mobile devices. For greater security the customer can store files with a password so that during file access the password will be required. This is a simple file protection that can be easily overcome by a hacker as there are tools available to unlock files easily. This is important for small and medium sized businesses to realize without having a false sense of security. Since Dropbox uses AWS' S3 service for file storage, it is not easy for a hacker to access the stored file.

On the SLA front many businesses are concerned with the lack of any service level guarantee from Dropbox. Moreover, as mentioned earlier, Dropbox does not own its file storage infrastructure as it is handled by AWS S3. The only security feature that Dropbox claims is that all data in transit are encrypted. This is one major reason why several businesses prefer other storage service vendors like AWS. Dropbox file access controls are much simpler compared to the traditional FTP approach. With the link provided by the file owner one can directly access the file without any additional authentication.

4.2.9   Box

One competitor to Dropbox that offers comparable service with a business focus is Box. This company was launched in 2007. Its service differentiation comes from dynamic, flexible, sharable content in any format on any device. Its approach to integration with best-of-breed applications such as Google Apps and Salesforce makes it more useful for businesses. The content collaboration feature appeals to business customers who range from small businesses to Fortune 500 companies. At the end of 2013, Box had over 8 million users.

Box guarantees a three 9s (99.9 %) uptime for service availability and provides greater security for content both in storage and transit. It uses strong SSL encryption for data in transit and data at rest are encrypted according to the AES 256-bit en-

Table 4.1 Cost comparison for cloud storage

Service provider	Free storage limit	Additional Storage Cost/Year	Remarks
Dropbox	2 GB	100 GB—$ 100 200 GB—$ 200 500 GB—$ 500	Could be increased up to 18 GB free by referrals at 500 MB/referral
Google drive	15 GB	100 GB—$ 60 200 GB—$ 100 400 GB—$ 240	Pure storage service without additional synchronization and sharing features of Dropbox
Microsoft SkyDrive	7 GB	50 GB—$ 25 100 GB—$ 50	Pure storage service
Apple iCloud	5 GB	25 GB—$ 40 55 GB—$ 100	Storage limited to customer owned files. No content purchased through iTunes store could be stored.
Box	10 GB	100 GB—$ 60/user 1 TB—180/user	File upload size for free service is limited to 250 MB and 2 GB and up for paid service

cryption standard. Box service has the SSAE 16 Type II Audit certification to assure businesses of the high level of compliance and controls. Box provides role-based access control for file access. Box's file sharing service is an efficient alternative to the traditional FTP method to share files. Businesses needing file sharing feature with their customers simply send the link to the file without the need for any additional authentication using login and password. Table 4.1 gives a cost comparison for file storage among the major services available now.

Even though Box does not have the user base of other cloud storage services such as AWS S3 and Windows Azure Storage, it has the ease of use and collaboration features that businesses are looking for in the cloud service. Box supports REST (Representational State Transfer) protocol for networked applications as well as SOAP (Simple Object Access Protocol) which is more complex than REST. In Table 4.2 we provide a snapshot of how Box compares with AWS S3 and Windows Azure Storage.

4.3 Niche Cloud Service Providers

In the previous section we considered several major cloud service providers. In the marketplace we have several other cloud service providers who are meeting the business needs for cloud service using the same benefits of cloud service as the main reason. In this section we will discuss in detail several niche cloud service providers who focus their attention based on a language, geographic region or specific industry. In order to put into context the services of niche cloud service providers we consider Michael Porter's famous Competitive Advantage generic strategies of 1980s. It consists of cost leadership, differentiation, and focus (Porter 1998). These three strategies have stood the test of time. Companies succeed when they focus on one or more of these three strategies. Some of the cloud service providers

Table 4.2 Comparison of Box, AWS and Azure for Storage

Service feature	Box	AWS	Azure
Protocols supported	REST, SOAP	REST, SOAP	REST
Encryption type in use	AES-256 for Enterprise users	AES-256	None by default. Customer chooses what is needed
Reports available for customer use for compliance purpose	Extensive control options for customer to setup	Customer can setup gathering access log	None by default
File size limit	2 GB for Business 5 GB for Enterprise	5 TB	200 GB block blob 1 TB page blob
Pricing method	Per user	Per month	Per month
SLA for availability	99.9 %	99.99 %	99.9 %
Service differentiation	1. Simple user interface 2. Integration with popular applications	1. File versioning 2. Set policy for content expiration	1. Integration with other Azure services

have realized the importance of these three strategies. The niche cloud service providers mainly focus on service differentiation while others concentrate on a focus area such as health care, automotive or finance. Our analysis of the niche providers shows that none of them are concentrating on cost leadership as a strategy.

The leader among niche cloud service providers is Salesforce.com. This company was launched in 1999 to provide Customer Relationship Management (CRM) service over the web. During the past 15 years this service has grown to be worth several billion dollars. Today, Salesforce.com provides the traditional CRM service and a business social network service called Chatter. The Chatter service is used by many professionals, similar to LinkedIn. The CRM service is offered as a SaaS service through Salesforce.com. The service is aimed at all types of businesses—small, medium and large. It offers four tiers of service—group, professional, enterprise and performance. All services are priced per user per month. The cost for these four tiers is $ 25, $ 65, $ 125 and $ 300. As part of the CRM service Salesforce provides the customers with leads the business can contact and maintain the contact once it is established. It supports the customer's applications to contact the potential lead through any of the preferred methods—email, mobile, social or the web. The target marketing is at the 1–1 level which makes the prospect to pay attention to the sales pitch. The application is entirely run on the Salesforce site and as such there is no client software installation or maintenance. Salesforce does not offer any uptime guarantees and no SLA. Instead, it offers End User License Agreement (EULA) which specifies how the application would run. As a SaaS and PaaS vendor, Salesforce focuses on having the applications run efficiently for the customer. It is to be noted that Salesforce experienced service outages over the past 5 years which made their services unavailable to customers at times. Their goal is to cut down on such outages and increase the ability to recover from any application failures quickly.

IBM's Smart Cloud Enterprise (SCE) is a niche cloud service focused on Infrastructure as a Service. It is geared towards testing and development environment. Since the acquisition of SoftLayer, IBM's Smart Cloud Enterprise service is being phased out in early 2014 because the SoftLayer environment provides highly automated self-service infrastructure on demand. SoftLayer is focused on bare metal servers. Bare metal servers provide a higher level of freedom for the businesses in choosing their virtual servers. Typical virtual servers operate within a host operating system environment and the virtual server resides within that host environment, leaving the management of resources within the hardware to the host operating system. With bare metal servers the customer has the ability to select their host operating system environment for their virtual servers. SoftLayer has over 200,000 servers in 7 data centers located in U.S., Europe and Asia. Its largest data center is located in Dallas, Texas. It's most recent data center addition was in Asia via colocation at the Digital Realty Trust, one of the largest data center providers. The main reason for this expansion was to reduce the latency time for gaming applications in the Asia region.

SoftLayer's strength is in provisioning bare metal servers in an automated manner. A bare metal server has all the computer hardware in it but lacks a preloaded operating system. The customer is able to build up the bare metal server to their specification. The main advantage of the bare metal server is that it still supports server virtualization but through the hardware rather than through a hypervisor layer on the host operating system. The latter and more traditional approach causes additional latency, hence the popularity of bare metal servers for applications that need higher performance speed. Niche service providers like SoftLayer preconfigure some of their bare metal servers with the types of virtual servers customers would need which make it possible for the customer to rapidly deploy their initial configuration of bare metal servers and add additional specialized bare metal servers shortly thereafter. This significantly reduces the launch time which in the traditional cloud service model would take several days or weeks to deploy. One other large provider of bare metal servers similar to SoftLayer is Internap. These niche providers offer innovative services by leveraging technology (Frost 2013).

The five areas of concentration for succeeding as a niche cloud service provider are localization of service, geographic location of service, specific industry focus, language focus, and ease of use with applications. We will address the specific industry focus in health care, automotive and financial sectors where cloud service providers have developed a niche. In the health care area one of the niche cloud service is Optum Cloud. It enables users to launch new health care applications quickly and inexpensively. It also enables users to communicate health data in a secure, HIPAA complaint manner. The Optum Cloud facilitates communication between health care professionals, hospitals, patients, insurance companies and public health officials. With the availability of health related data in one location for a specific geographic region or city the Optum Cloud facilitates cohort analysis and benchmark development for service standards. It also enables health service providers to share patient data in a secure manner. There are several other niche cloud service providers who focus on health care industry such as CareCloud and Dell Secure Healthcare Cloud.

In the case of the automotive sector two niche service providers are IBM and HCL Technologies. The automotive industry is preparing for meeting customer expectations in automobiles. The automobiles are viewed not only as transporting people but also keeping people connected and entertained as they use the automobile. IBM's Smarter Planet initiative is geared towards this goal (IBM 2012). HCL Technologies is based in Asia and through its Agora service it is aiming to meet the needs of the automotive companies in building cars that will meet the user expectations which include high level of connectivity and location based services. Agora is a service management platform for cloud service. HCL's investment in providing cloud based service for the automotive sector is worth $ 6.5 billion (HCL 2014).

Another industry vertical that has a significant need for cloud based service is the financial industry. Two major companies providing niche cloud service in this sector are Sapient and Computer Services Corporation (CSC). Financial service industry is extremely data intensive and currency of data is very critical for decision making. The cloud service should be capable of addressing risk calculation, performance attribution and trade reconciliation quickly for financial services. The industry needs the ability to introduce new services rapidly. When employees are distributed globally it is imperative for the service to have assured security. This can be achieved using Virtual Desktop. Cloud service enables making Virtual Desktops available globally within a few hours. Services offered by CSC and Sapient support the needs of the financial services industry.

Most of the cloud service providers are U.S. based. However, there are several niche cloud service providers around the world who focus on a particular geographic region or provide service in a local language. We describe some such services from Europe, Australia and Asia. It is worth noting that 88 % of all cloud service providers are U.S.-based and only 6 % of the cloud service providers are Europe-based. Some of the European cloud service providers are Dimension Data and Colt Enterprise Cloud. Both these companies provide cloud IaaS service. Cloud Central and BitCloud are two of the cloud service providers from Australia. Cloud Central offers enterprise level cloud service as well as a Government Cloud service for government agencies. BitCloud offers managed cloud services and enterprise level cloud computing services. In Asia, CloudStar Asia provides cloud services for the enterprise as well as small and medium sized businesses. Trend Micro is based in Japan and offers cloud security services. HCL Technologies is based in India and offers cloud service for the automotive sector. Acens Technologies is a cloud service provider from Spain offering cloud service in Spanish language.

4.4 Third Party Facilitators

Cloud computing providers are numerous and the consumer seeking a cloud service may be at a disadvantage to look into all aspects of the cloud service provider's claims in selecting a suitable cloud service provider. Third party facilitators help the customers choose the appropriate cloud service and also help the customer migrate to a different service when the service provider is not able to meet the customer expectations either from the SLA perspective or from the resources availability. Such third party services are labeled Cloud Service Brokerages (CSBs). The CSBs help the customer deploy and integrate apps on multiple clouds as well as help with cost saving options in service selection because the cloud uses the pay-as-you-go model.

Some CSBs provide migration, VM portability and API management. According to Gartner (Gartner Research 2009) the CSB role could involve:

1. Aggregation
2. Cloud Service Intermediation
3. Cloud Arbitrage

We will briefly expand on these concepts and identify the benefits to the customers in using the CSB service.

Aggregation involves combining multiple services to provide an output that is value added to the customer. These specialized services focus in aggregation and as such provide a much needed service by small and medium sized businesses that do not have such capability. The aggregators take two related services and aggregate them to add value to the customer. For example, a real estate search could yield some properties but combining that search result with a mapping product places the search result in a location map. The resulting output adds value to the customer. Two cloud service aggregators are Boomi and Cast Iron. Cloud Service Intermediation means the intermediary is able to provide value added service to the customer such as identity and access management. This is the common form of brokerage service. Intermediation brokers also help the customer by verifying the pricing and billing for cloud services. Alexa Web Information service is one such intermediation broker. The intermediaries contract for services such as identity management with major vendors such as AT&T or Verizon. Cloud Arbitrage is similar to aggregation except that the services aggregated are not fixed always. For example, the CSB could select a specific email service for the customer from among the multiple email services provided by the cloud service provider (Plummer 2009). Another example of arbitrage is when the CSB provides the backup service for a customer or performance guarantee. This is an evolving service. The need for such a service exists based on the existing brokerage services outside of cloud service. There are independent insurers who compare the insurance rates for a customer from among several major insurers and find the best rate. This example takes into account how different providers assess risk and offer their rates. Such comparative service is common in the insurance industry for various types of insurance products. One example of a CSB in general is Akamai. It provides validation service by checking all incoming web connections to whitehouse.gov web site from being attacked as part of a Denial of Service attack. Another example of a CSB is Appirio from San Francisco that provides brokerage service to customers.

Cloud Service Brokerage service is a major service industry. According to Gartner Research 20 % of all cloud services used will be through a cloud broker by 2015, which is a significant increase from the current 5 % use of cloud services through brokers. The brokerage service is sought after by many potential cloud customers as they need help in navigating the many details in selecting and benefiting from the various services being offered by a cloud service provider. It is projected that the annual IT spending on CSB services will reach $ 100 billion in 2014. This type of intermediary service is not unique when the type of interaction involves complex matter. To leverage the coordination service needed by cloud customers for their multiple cloud services the Australian company Cloud Manager has created a new cloud brokerage service called CloudMgr that provides an integrated cloud management solution. It not only integrates the various services used through AWS but also the other cloud services the business may use as well as with leading business productivity services such as Autotask. Surveys have shown that even though cloud computing is growing at a Compound Annual Growth Rate (CAGR) of 25%, the overall cloud computing is small relative to the IT outsourcing market. According to a 2013 Interxion survey in Europe, the traditional hosting (managed hosting and dedicated hosting) account for nearly 82% and cloud computing accounts for 18% share of the IT services market (Interxion 2013). We conclude this section with two typical examples from other industries where such intermediaries play an important role: real estate dealings and financial advising.

4.5   Emerging Cloud Standards

Cloud computing, even though it is growing rapidly, is lacking a global standard that would let customers evaluate the services better. In this section we will discuss three main efforts underway to develop standards. These are: OpenStack, OpenNebula and Eucalyptus.

OpenStack is an open source software for building public and private clouds. It is a massively scalable cloud operating system (OpenStack 2013). It was developed by Rackspace and NASA. The developer and user base for OpenStack spans over 130 countries. Once many cloud service providers adopt the OpenStack operating system then customers will have interoperable service giving greater comparison of services. Moreover, customers will be able to change service providers easily when their needs change. Among the major cloud service providers, Rackspace and HP use this technology. However, their implementations of the OpenStack differ and their offerings are not fully interoperable. CERN in Switzerland has adopted OpenStack for building a private cloud for the nearly 11,000 physicists around the world. This private cloud is expected to grow to 150,000 virtual machines by 2015. Some of the major users of OpenStack are AT&T, HP, Rackspace, IBM, RedHat, Ubuntu, Suse, Cisco, Dell, Intel, Ericsson, Hitachi, Yahoo, EMC^2, Comcast, PayPal, NEC, Alcatel-Lucent and Fujitsu.

OpenNebula is an open source enterprise cloud service model. It is meant for many of the Linux distributions. It is an industry standard for data center virtualization. It can be customized to fit into any data center (OpenNebula 2013). Since VMware dominates the virtualization field its vCloud service competes with OpenNebula. This standard has interfaces for the popular AWS Elastic Cloud Computer (EC2) and Elastic Block Storage (EBS) instances. Since cloud service uses the pay-as-you-go model this standard has a fine grained accounting system that integrates with all billing systems for the chargeback feature of cloud service. OpenNebula standard is platform independent. It supports public, private and hybrid clouds. As an open source software it is released under the Apache license. It supports security very well through the use of fine grained Access Control Lists to limit access to content. Some of the major users of OpenNebula are IBM, Dell, NASA, Unisys,

China Mobile, Akamai, Blackberry, Harvard University, Academia Sinica, KPMG and CentOS.

Eucalyptus is an open source software platform for IaaS service in private or hybrid cloud environments. Eucalyptus is an acronym for Elastic Utility Computing Architecture for Linking Your Programs to Useful Systems. It was formed from a research project at the University of California—Santa Barbara. Eucalyptus Systems was formed as a for-profit organization in 2009 in spite of the open source software goal. It supports virtual machines using Windows or Linux operating systems. It integrates well with AWS's EC2 and S3. It has a cooperate-compete relationship with AWS in which it touts easy move back and forth between AWS' public cloud and Eucalyptus' private cloud (Nurmi 2009). Eucalyptus effort in contributing a cloud standard is minimal except through the open source software it has developed. Its status as a for-profit organization does not bode well with its stated goal of being an open-source software. It has joined Dell as a technology partner in providing a cloud-in-a-box software for developing a private IaaS cloud.

4.6 Summary

In this chapter we have reviewed the cloud services offered by a number of major service providers. From this analysis it became clear that each cloud service provider has certain strengths in the type of service that they provide. All these cloud services enable organizational IT to support business growth and innovation. Moreover, they facilitate efficiency in IT service delivery and are agile enough to meet the business needs. Besides the major cloud service providers there are also several niche service providers who focus their service for specific geographic regions or an industry such as health care or a language. Most of the major cloud service providers are located in U.S. and some of the niche service providers are located in various countries. The niche service providers are also able to meet certain legal requirements of data location and protection within their country borders. Moreover, we considered the role of Cloud Service Brokers and their important contribution to customers in selecting and using a particular cloud service. The Cloud Service Brokers offer a variety of services that are value added and so the cloud customers are able to benefit. We noted that is a large industry within cloud services. Our discussion included the importance of global standards for cloud services and how OpenStack, OpenNebula and Eucalyptus are contributing to the maturity of cloud service in a standardized way.

4.7 Review Questions

1. Identify how computing growth is impacted by the lack of global standards. How is the cloud computing industry trying to address the standards aspect?
2. Describe the services offered by the top tier cloud service providers Amazon Web Services, Microsoft and Google and their benefits and drawbacks.

3. Compare and contrast the services of Rackspace, Verizon Terremark, Salesforce and HP with respect to the benefits and concerns for customers.
4. Describe the services offered by Dropbox and Box and how they are changing the way businesses use cloud storage. Compare the relative benefits of using either service.
5. Describe how the cloud services Salesforce and Softlayer are meeting the differing needs of cloud customers.
6. Describe the role of Cloud Service Brokers in the current state of the industry and how they are meeting the service integration needs.
7. How is the cloud computing industry striving to address the lack of global standards through industry consortia? Explain.

References

Cloud Industry Forum. (2014). Overview. http://www.cloudindustryforum.org/. Accessed 5 Feb 2014.
Cloud Security Alliance. (2014). Alliance Overview. https://cloudsecurityalliance.org/. Accessed 5 Feb 2014.
CSC. (2014). Cloud Computing for Financial Services. http://www.csc.com/financial_services. Accessed 5 Feb 2014.
Frost. (2013). Leveraging technology for innovation and success, Frost & Sullivan White Paper.
Gartner Research. (2009). Three types of cloud brokerages will enhance cloud services.
Gartner Research. (2013). Magic quadrant for cloud infrastructure as a service.
HCL. (2014). Cloud based services. http://www.hcltech.com/automotive/cloud-based-services. Accessed 5 Feb 2014.
IBM. (2012). Cloud computing for automotive. http://public.dhe.ibm.com/common/ssi/ecm/en/giw03003usen/GIW03003USEN.PDF. Accessed 5 Feb 2014.
IDC. (2013). IDC cloud research. http://www.idc.com/prodserv/idc_cloud.jsp. Accessed 3 Jan 2014.
Interxion. (2013). The evolution of the European cloud market, Whitepaper.
Microsoft. (2010). Forecast: An improved economy in the cloud—An introduction to cloud computing in government. White Paper.
Nurmi, D. et al. (2009). The Eucalyptus Open-source Cloud Computing System. 9th Annual IEEE/ACM Symposium on Cluster, Cloud and Grid Computing, Shanghai, China.
Open Data Center Alliance. (2014). Overview. http://www.opendatacenteralliance.org/. Accessed 5 Feb 2014.
OpenNebula. (2013). http://opennebula.org/. Accessed 12 Jan 2014.
OpenStack. (2013). http://www.openstack.org/. Accessed 12 Jan 2014
Plummer, D., & Keeney, L. (2009). Three types of cloud brokerages will enhance cloud services. Gartner Research Report.
Porter, M. (1998). *Competitive advantage: creating and sustaining superior performance*. New York: Free Press.
U.S.-EU Safe Harbor Agreement. (2000). U.S.-EU Safe Harbor Overview. http://export.gov/safeharbor/eu/eg_main_018476.asp. Accessed 12 Jan 2014.
U.S.-Swiss Safe Harbor Agreement. (2008). U.S.-Swiss Safe Harbor Overview. http://export.gov/safeharbor/swiss/eg_main_018519.asp. Accessed 12 Jan 2014.
Verizon. (2013). 2013 State of the cloud report, White Paper.

Chapter 5
Cloud Computing Security

Abstract Security aspects of cloud computing draw much attention. Many cloud customers feel that their lack of control over hardware and software makes their information vulnerable for compromise on the cloud. The security issues surrounding the cloud vary among the different types of cloud services such as SaaS, PaaS and IaaS. Among the cloud deployment models only the public cloud has several vulnerabilities. Businesses feel that since they do not control the cloud infrastructure any data stored in the cloud is insecure. It is more a perception issue than something that is inherently insecure. The cloud service providers are trying to reassure the public of their security practices and provide third party audits to back up their claims. Moreover, all the major service providers seek the enhanced SSAE 16 Type II Audit and the ISAE 3402 international reporting standards compliance certification. In this chapter we will analyze the security implications for businesses from the perspective of compliance with laws and industry standards as well as certifications carried by the service provider. Moreover, the service providers facilitate implementing both access control mechanisms and organizational control policies to limit the number of privileged users with access to customer data. Also, we discuss the proactive steps an organization could take to protect their data in transit and storage.

Keywords Security · Compliance · Certification · Access control · Organizational control · Data center

5.1 Introduction

Ever since cloud computing was launched nearly 8 years ago businesses have embraced this new technology tool with some reservations. The benefits offered by cloud computing are too numerous to ignore, especially when it helps small and medium sized businesses to obtain computing services at an affordable cost. From a business perspective, moving the cost for technology management from capital expenditure to operational expenditure is preferred by organizations. Furthermore, these types of organizations lack the necessary computing expertise to manage an information system. Large organizations have the resources to manage an information system on their own but given the rapid advancements in technology and the

S. Srinivasan, *Cloud Computing Basics,* SpringerBriefs in Electrical and
Computer Engineering, DOI 10.1007/978-1-4614-7699-3_5,
© Springer Science+Business Media New York 2014

constant need to make available the information on mobile devices they find the cost factor daunting. Thus, cloud computing offers an effective alternative to all types of businesses to have a reliable and affordable computing system since the users pay only for the services that they use.

Businesses are accustomed to controlling their computing infrastructure and that gave them the sense of security that they needed. With cloud computing they had to cede control of the physical infrastructure to cloud service providers located in remote locations, with their data and computing resources distributed across a wide geographical area. From the user perspective this lack of control over the infrastructure is unsettling to several businesses. From a practical perspective the small and medium sized businesses would not be able to provide the level of physical security for computing infrastructure that a large cloud provider could offer. On this aspect most businesses agree that a cloud service provider's assurances are adequate. In spite of that the concern for businesses is that they are concerned with the need to share the same physical infrastructure with other businesses due to virtualization. The cloud service provider is able to offer "unlimited" computing resources on demand using the virtualization concept. This way they can have a higher server utilization rate and pass on the cost savings to the customers. When companies own their servers the server utilization rate is usually less than 20 % according to multiple studies (McKinsey 2008; Gartner 2009). Moreover, companies are paying attention to energy savings and are supporting green energy initiatives. According to CDW's Energy Efficient IT Report developed from a survey of 760 people representing public, private and government sectors shows that cloud computing saves energy in two ways. First, the use of virtualized servers and storage accounts for 28 % energy savings. Second, there is a significant amount of savings that is not quantified but verifiable in that personnel can access cloud computing from remote locations, saving the commuting and office space cost. In this regard some companies look for service providers with energy saving technology in place (CDW 2012).

Cloud security involves the type of cloud service and deployment model chosen. In order to understand these differences let us briefly consider these service types and deployment models. The three service types are Software as a Service (SaaS), Platform as a Service (PaaS) and Infrastructure as a Service (IaaS). Collectively, these are referred to as the SPI (SaaS, PaaS, IaaS) model. SaaS is the most heavily used service type and cloud computing is usually associated with this type of service. It is popular with small and medium sized businesses. It accounts for 65 % of the cloud services. In this service the cloud service provider controls all aspects of the service provisioning. Customer concern is due to possible data leakage because of the multi-tenancy of different businesses using the same physical server due to the virtualization capabilities (Sengupta et al. 2011). In PaaS service the customer chooses the computing platform they need. This service is typically used by developers for testing their application under several platform scenarios such as Windows, Mac or Linux. It accounts for 15 % of the cloud services. The security concern is similar to SaaS regarding multi-tenancy and data leakage. IaaS gives the customer maximum freedom as to the choice of operating system or applications that they run. Typically IaaS is used by large businesses as an extension of their internal information system. This service accounts for 20% of all cloud services. In this case the customer has higher levels of control and so they are responsible for application security aspects. In all these three types of services the physical security is still the responsibility of the service provider. Major service providers offering these types of services are Amazon Web Services (AWS), Microsoft, Google, Verizon Terremark, Rackspace, Salesforce and Dropbox. AWS' major offerings are Elastic Compute Cloud (EC2) and Simple Storage Service (S3). Google's cloud service consists of Google Apps which includes Gmail, Google Docs and Google Drive. Microsoft cloud services include the Windows Azure platform and the Office 365. Verizon Terremark's cloud offering is vCloud. Rackspace's cloud service consists of both PaaS and IaaS. Salesforce's primary cloud offerings are its most popular Customer Relations Management (CRM) application and the associated Chatter social network service. Chatter is similar to LinkedIn in scope. Dropbox's cloud offering consists of cloud storage.

Cloud storage is one of the important services that cloud service providers offer. This type of service evolved in two different strands. In one strand, AWS offered storage service as an appendage to their cloud computing service and it usually involves large storage volume. In the other strand, Dropbox pioneered small amounts of file storage for individual users for sharing content over the web. This latter service caught on with users worldwide and Dropbox now has over 200 million users globally, with a daily storage of over 1 billion files. Because of this amount of storage space needed in an elastic manner, Dropbox uses AWS' S3 service for storage of all user files. This has become a fiercely competitive cloud service with Google and Microsoft actively competing for storage service. We summarize in Table 5.1 these cloud storage services, giving additional details on the volume of storage, cost, devices supported and the type of content planned.

The three common deployment models are public cloud, private cloud and hybrid cloud. The largest of these three deployment models is the public cloud and it mostly symbolizes cloud computing in general. Public cloud implies that all customers share physical servers with other customers because of virtualization capability. Most of the public cloud users are small and medium sized businesses, but large businesses also use some aspects of the public cloud such as storage. The public cloud accounts for 70% of the cloud service deployment models. It also has the most security concerns for the customers. Amazon Web Services is the largest public cloud service provider. By definition the private cloud signifies dedicated servers and so security is not a concern. At the same time private clouds are relatively more expensive than public clouds. The user should be able to all the resources available through a private cloud, which invariably means that it is suited for large businesses. Private clouds account for 24% of all cloud deployment models. Hybrid cloud implies a combination of both public and private clouds. Hybrid clouds are used by large businesses. Often a large enterprise that needs temporary resources to test some applications uses the public cloud taking advantage of the service elasticity and cost savings features of the public cloud. For all their mission critical applications they have their own internal computing resources or private clouds. Hybrid clouds account for 6% of all cloud deployment models.

Table 5.1 Summary of cloud storage services

Service provider	Storage level	Cost	Devices supported	File content support
AWS S3	5 GB Additional 1 GB	Free $ 1/yr	Windows and Mac devices	All file formats. Cloud music player
Google drive	9 GB 16 TB	Free $ 4,000/yr	Any device with web browser	All file formats. Music limited to user owned content
Microsoft windows live	25 GB	Free	Windows pc, Windows phone	Document files, photo. No music file support
iCloud	5 GB	Free	iOS devices and Mac	Music, photo and document files, high level of content synchronization on all supported devices
Dropbox	2 GB 100 GB	Free $ 200/yr	Windows and Mac devices	All file types. Built-in audio player

Cloud service providers manage their operations using several data centers distributed globally. These data centers serve dual purposes. First, it enables the service provider to have distributed service provisioning which helps with service redundancy. Second, by locating data centers globally the chances of having a data center close to a service customer is high. This is important for customers in many countries because they may have a compliance requirement that the data may not leave the region. Facilitating a global standard for the operation of data centers is the Telecommunications Industry Association Data Center Standard TIA-942 which is used widely. The TIA-942 standard promotes energy efficiency as well (TIA 2012). One measure for adoption of a particular cloud service provider is the location of a data center closer to the customer. In light of recent revelations of NSA surveillance and the force of the USA PATRIOT Act, many European governments and businesses are concerned about storing data in US servers. However, such data is protected for privacy under the US-European Safe Harbor Agreements. Companies like Salesforce have counted on the Safe Harbor Agreements for European customers. However, to overcome any hesitancy on the part of some businesses and governments many cloud service providers are opening up new data centers in Europe. In Table 5.2 we summarize the number of data centers that each cloud service provider has globally.

Security is critical for most organizations and consumers. We noted already that cloud computing has many benefits to offer the consumers. In certain settings the benefits far outweigh the security concerns. We point this out to show that security has to be balanced with the overall objective of the user. In a rural healthcare setting where cloud computing is used to access the health data of ordinary citizens because of lack of established infrastructure everywhere, the citizens who are patients are more interested in getting the healthcare they need rather than worry about the possible security violations in handling their health data. For these patients living with less suffering takes higher precedence over the privacy protection needed for their health data. In this chapter we will consider several security best practices,

Table 5.2 Summary of global data centers for major cloud service providers

Service provider	North America	Europe	Asia	Latin America	Australia
Amazon web services	8 major cities in US	UK Germany Ireland Netherlands	China Japan Singapore	Brazil	
Google	6 cities in US	Belgium Ireland Finland	Taiwan Singapore	Chile	
Microsoft	6 states in US Canada Mexico Puerto Rico	20 European countries such as UK, France, Germany, Italy, Spain	China Singapore Japan South Korea India Malaysia Taiwan	Brazil Chile Colombia	Australia New Zealand
Verizon Terremark	21 major cities in US Canada	UK Spain Netherlands Belgium		Brazil Colombia	
Rackspace	3 major cities in US	UK	China		Sydney
Salesforce	2 centers each in east coast, west coast and Midwest in US	UK[a]	Japan Singapore		
IBM	Major cities and financial hubs in US Canada Mexico	UK Germany	China Japan India	Brazil	Sydney Melbourne Brisbane Adelaide

Microsoft has an additional data center in Israel. Terremark has an additional data center in Turkey
[a] Operational in 2014

requirements from governments and industry for compliance with regulations and standards, and the controls in place.

The Big Four cloud service providers are Amazon, Google, Microsoft and Salesforce. Their combined market share exceeds 80 % of all cloud services. It is important to note that the Big Four are focused more on low-risk services that cater to small and medium sized businesses more than large enterprises. For example, when a large enterprise wanted to extend their company email to several thousand employees on their own devices, Microsoft could only commit for 99.9 % availability. Because of the success of their offerings to a very large group of global users the Big Four are not contributing to growing the security practices. The National Institute of Standards and Technology (NIST) working on guidelines for security best practices was unable to get the needed details from Salesforce, which does not disclose its security practices (NIST 2011). Of the Big Four, only Amazon has a packet filtering firewall that the customer can control on the public cloud. At the same time large telcos such as AT&T, Verizon and British Telecom are entering the cloud service market geared towards the enterprise level businesses. In order to provide the necessary security assurances, a new business called Appirio has emerged that offers document management security features for Amazon, Google and Salesforce clouds. These trends show that cloud services are expected to attract more providers with specialized focus. Like all other security controls in the marketplace for organizational infrastructure, cloud service also has security controls (Hashizume et al. 2013). Unlike the infrastructure of a single organization, the cloud attracts attackers because they realize that they will have greater impact since the user base is large. The complexities in security policies increase as well because of the large number of users. Cloud industry has to balance the protection mechanisms used versus the security constraints.

5.2   Security Best Practices

Security is a double edged sword. From one perspective people and businesses look at the security features as a savior. It helps protect the confidential information of the users. From another perspective it is a hindrance to rapid business growth. In the case of cloud computing since customers already feel the pressure of lack of control they tend to focus more on the security aspects of information storage and retrieval. In this section we will focus on several best practices that are available for the cloud service providers to adopt and also how some vulnerabilities escape service provider attention.

One of the important considerations for a cloud customer is the reliability of the cloud service. This is measured in terms of the service uptime of the provider. This means that the service provider is able to guarantee its service availability a through Service Level Agreement (SLA) at a certain rate. The unit of measure for this is the number of 9s in the service level. Major cloud service providers such as Amazon Web Services (AWS), Google, Microsoft, Rackspace and Salesforce all provide service guarantee at three 9s level, i.e., 99.9 % availability. For normal computing resource availability this is a high level of uptime. However, when the service offerings are concentrated with a few providers then a large number of people depend on those services and so the customer expectation is 100 % availability. For example, Google's Gmail service is used by over 450 million users globally. Gmail is a cloud application available for free for individual users. When the Gmail service is interrupted even for a few minutes it has major ramifications globally. So, one of the best practices is to make the cloud service reliable. Even a three 9s uptime guarantee allows only a total downtime of 9 h per year. An uptime guarantee of four 9s allows for only a 53 min total downtime per year (Srinivasan 2014a). A recommended best practice would be to make the cloud offering reliable with at least a three 9s uptime. Customers should be vary of any uptime guarantee at any higher level as they would be very difficult to keep. In reality all major cloud service providers have experienced significant outages over the past 5 years that would violate the uptime guarantee. In these situations the service providers have offered cost credits for users. It is important for the service providers to explain to the customers the reason for the outage and how they would prevent a recurrence. This is expected of service providers to maintain customer trust.

Cloud service providers would not know the sensitivity nature of the data stored by customers. For this reason the service providers feel that they are not responsible for protection of customer data (Ponemon Institute 2011). Customers on the other hand expect the service provider to protect their data since they are responsible for managing the computing infrastructure. In order to satisfy customer expectations the cloud service provider should make their service management aspects transparent. This includes making available the list of management personnel with elevated privileges to access the server data. Moreover, there should be a log of all data access by management personnel that the customer can access on demand. This will provide the customer the necessary confidence about the data management practices (Rittinghouse and Ransome 2009).

Confidence engendered by service provider security practices is often a matter of perception by the customer. To enhance such perception the service provider should make available their security policies and relevant data to third party registries. One such registry is maintained by Cloud Security Alliance (CSA), an industry consortium that develops security standards and best practices (Cloud Security Alliance 2013). CSA's members include all the major cloud service providers— Amazon Web Services, Google, Microsoft, Rackspace, Salesforce, IBM and Cisco. CSA registry is called STAR. It stands for Security, Trust and Assurance Registry (CSA STAR 2011). STAR collects a variety of security practices data such as physical security controls, number of privileged users at the management, service provider audit data and compliance certifications. CSA makes available a security questionnaire titled Consensus Assessment Initiative Questionnaire (CAIQ). Responses found in this questionnaire enable the potential customer to assess the cloud service provider attributes such as compliance with standards like HIPAA, security and governance policies. The third tool available is the Cloud Control Matrix (CCM) from CSA. The Cloud Control Matrix has an extensive set of questions for which the service provider's data helps the customer assess the security. For example, the CCM contains data about the business' compliance with industry-accepted security standards, control frameworks, audits and regulations. Moreover, the service provider reports in CCM all data concerning risk management, business continuity, data leakage prevention, automated equipment identification for connection authentication and data access security for mobile devices. The fourth tool available for the service provider is the Cloud Commons' Service Measurement Index (SMI). The SMI provides a set of business-relevant Key Performance Indicators (KPIs) (Cloud Commons 2011). All these help enhance the reputation of the service provider through the use of commonly accepted security practices The National Institute of Standards and Technology maintains a vulnerability database that could also be consulted by customers to assess the security aspects of the service provider (NIST 2013).

Use of sound security practices enables the cloud service provider to gain customer trust, especially when services are provided using remote infrastructure. Cloud service providers understand that customer trust is earned and cannot be taken for granted. One way to earn customer trust initially is to have an association with other trusted entities. For example, major multinational corporations such as Amazon Web Services, Microsoft, Google, Apple and Salesforce have earned customer trust over many years of reliable service. Even these providers partner with other best-of-breed companies such as Cisco and RSA for specific services like firewall control and encryption. Customers place trust in organizations that have a good reputation. Thus, reputation of the organization is one aspect of trust.

Cloud services should use new methods to authenticate users. In this context we will address two potential vulnerabilities in authentication using existing userid-password combinations. This technology was developed over 50 years ago when computing resources were not advanced as they are today. Service providers may not use encryption technology to store passwords in the first place and where encryption is use it may not be strong. Systems that do not use encryption run the highest risk. To identify service providers who do not use encryption to store passwords, simply use the 'Forgot Password' feature to recover the password. If the password comes back in clear then that provider's system has a vulnerability. Service providers who do not require strong passwords containing special characters are likely using databases to store the passwords where the special characters have special meanings. Without the use of strong encryption, user authentication in cloud services is not adequate. Today's personal computers have many times the processing power of large computers of 30 years ago. Consequently, cracking an encryption using brute force is possible with simple encryptions. Because of such vulnerabilities breaches occurred at LinkedIn and eHarmony, two widely used cloud services. To overcome the encryption problem a simple method exists which adds an additional value to the string, called salt, that varies from password to password. The use of salt with passwords negates the ability of the use of rainbow tables for fast lookup to find password matches. Another method to enhance user authentication in the cloud is to use a two-factor authentication. The two-factor authentication is a requirement in US for financial transactions over the web. In this approach the user id-password combination is supplemented with unique security questions associated with the user. Financial institutions use an added level of security with two-factor authentication with a security image chosen by the user for their login. These types of authentications are application-based and so the customer should verify that the cloud service is capable of running such applications for authentication.

Cloud service security should extend beyond cloud processing to cloud storage as well. The basic requirement of any cloud storage is the protection of data confidentiality, integrity and availability. It is typical for businesses to store data on the cloud when it is not heavily used. Healthcare industry generates large volumes of data and they get stored on the cloud using the HIPAA mandate of data anonymization. Theoretically such data should protect individual identity. However, using publicly available data about individuals a test was run against the Massachusetts Group Insurance Commission data stored in the cloud. A search of this database using some common data such as zip code, gender and date of birth revealed the medical history of the State Governor (Chow et al. 2009). This is an application vulnerability and it clearly shows that not all cloud service providers may have the tools to check the application vulnerabilities when they offer such services as part of Software as a Service.

Cloud service benefits from virtualization, which means that on a physical hardware one could emulate multiple hardware running different operating systems. VMware is the leader in offering virtualization software. When a physical server consists of multiple virtual machines (VMs), each VM runs in a Virtual Machine Environment (VME). VMEs are popular with researchers dealing with malicious software detection. Carpenter, Liston and Skoudis show in detail how malicious hackers detect the presence of VMEs and avoid hardware running VMware because they could not attack such machines easily (Carpenter et al. 2007). They point out further how they could hide the VMEs from attackers. This shows the importance of virtualization and at the same time how malicious hackers could detect VMEs and try to take advantage of known vulnerabilities in operating systems and escape to the host operating system from the guest operating system and cause serious damage. Along the same lines Zhang, Juels, Reiter and Ristenpart show that when VMs are used a malicious hacker could launch a side channel attack that lands the attacker in someone else's computing space from which the attacker could cause serious damage (Zhang et al. 2012). Even though VMs are inevitable in cloud environments these two security problems should make the customer aware of protection mechanisms needed against such attacks. A recent trend in cloud services is the use of "bare metal" servers instead of Virtual Machines. Bare metal servers function similar to Virtual Machines except that the customer who launches the bare metal server decides on the operating system to run on these machines (Softlayer 2013). One large company providing bare metal servers is Softlayer, an IBM-owned company.

Security services are difficult even in organization owned infrastructure. In cloud environment it is all the more difficult because the customer does not own the infrastructure. In this case the customer would benefit from the services of a third party cloud broker who would look for secure applications that could be run in the customer cloud area as well as monitor the security practices of the cloud service provider. Third party Cloud Services Brokerage (CSB) is a large business with many major companies providing this service. Some of these large companies are IBM, Microsoft, Dell, HP, Google, Salesforce and Rackspace. According to MarketsandMarkets forecast the CSB market is expected to grow from $ 1.57 billion in 2013 to $ 10.8 billion in 2018.

One aspect of dealing with cloud security involves risk management. As we mentioned earlier, cloud system reliability is one of the risk characteristics. All major cloud service providers have experienced service outages that have resulted in these companies providing service credit to the customers because of SLA requirements. Based on the reality of the situation rather than the SLA the cloud customer can better manage risk by using the services of more than one cloud service provider. For example, the customer could use one cloud service provider for all primary functions but use another service provider for cloud backup. This way, when the main service provider's system is not up then the customer could access their data from the other service provider where the backup data is stored. In order to do a total backup on a consistent basis one could adopt disk mirroring. Even though this is slightly expensive for the customer, at least it provides a risk mitigation when one service fails.

Risk management has four basic aspects. These are: Avoidance, Mitigation, Sharing, and Acceptance. Depending on the type of cloud service used the avoidance category has some strategies. With SaaS service used on a public cloud, which is the most used combination of cloud service, the customer has to use encryption technology to protect the data in transit and rest. This way, even if the data is lost by the service provider it would not cause harm to the customer. Risk mitigation involves knowing the risk and taking alternative steps. For example, when health data needs to be shared with a third party, it could be made available with a password and the password itself is transmitted using a different communication channel. Given the public nature of cloud services, sharing risk is one of the best options for many customers. This can be achieved via an explicit section in the Service Level Agreement. Another method is to use a third party cloud broker who could handle the same. For example, the customer is able to share their risk in storing data with the service provider. The service provider assures the customer that their data will not be released to a third party or government agencies without customer consent or notification, depending on the nature of the need for data disclosure. The final risk management strategy involves accepting that certain risks are inherent in doing business. For example, when a small or medium sized business contracts with a third party payroll service provider such as ADP, they accept the inherent risk that ADP could lose control of the customer data (Srinivasan 2014b).

Having identified some risk management strategies above we can recommend some solutions. First, much of the customer risk is centered around the use of public cloud, especially for placing any sensitive data on the cloud. The recommendation here would be to use a private cloud or institute a Virtual Private Network access to the data, which is kept in encrypted form (COSO 2012). The drawback to this solution is that if the data were to be accessed by mobile devices then they may not have the necessary computing power to decrypt the content retrieved. So, some restrictions to that effect may have to be instituted. Second, when using a cloud service provider data is in a central location and so some governments may try to access it without the knowledge of the data owner. The service provider should have policies against such access and make available data access logs to the customer on demand. Third, the cloud service provider should make available data in a central repository like the Cloud Alliance Security, Threat and Assurance Registry (CSA STAR 2011) that makes it easy to assess the risk of cloud computing. Fourth, the service providers should be required to notify customers in the event of a security breach. Fifth, service providers should report on incident handling and monitoring activity to build customer trust. Sixth, the service provider should provide assurances that data deletion is propagated through all backup copies. The European Union Report titled "Cloud Computing: Benefits, risks and recommendations for information security" has many such features identified (European Union Report 2009).

In concluding this section we note that cloud security has remained a major concern for many global businesses for many years according to several global surveys. Alcatel-Lucent's 2011 study of 3886 IT decision makers around the globe showed their second most concern with cloud services was data security. Gartner Research survey in 2009 and 2012 shows that cloud security is the top concern for enterprises planning to move their data into cloud (Blum 2009). The American Institute of CPAs' study in 2013 shows that security is the top concern for CPA firms moving to the cloud. Global Survey of internet infrastructure decision makers in 2013 by Internap Network Services clearly shows that security is the top concern of many companies that are not currently using cloud computing services. Associated with the security concerns is the frequent occurrence of data breaches at the central repositories such as the cloud service provider storage. These are a sampling of many surveys which show that security remains a top concern for existing cloud customers as well as potential customers.

5.3 Compliance and Certifications

In several industries such as health care and finance the businesses have to certify that they meet strict security guidelines that are mandated by laws. In this section we will review all the important compliance requirements and the possible certifications that a cloud service provider will be able to acquire. These compliance requirements and certifications assure the potential customer of the sound security practices of the cloud service provider. Any healthcare related business moving their computing operations to the cloud will have to certify that their computing service meets the Health Insurance Portability and Accountability Act (HIPAA) requirements. Patient health data protection is paramount under HIPAA. Cloud service provider will be able to get HIPAA certification upon meeting the strict security requirements concerning the infrastructure and management policies of the service provider.

Businesses associated with finance industry have to meet Sarbanes–Oxley Act (SOX) and Gramm–Leach–Bliley Act (GLBA) requirements concerning sound internal controls and privacy policies. Two of the important internal control requirements of SOX are separation of responsibility and the principle of least privileges. The cloud service provider having internal control policies that restrict the same user from authorizing an action and performing the action would meet the SOX requirement for separation of responsibility. Likewise, if management personnel are granted only those privileges that are required to perform their duties then it would meet the principle of least privileges. The GLBA requires the business to protect the privacy of the individual or business that they are dealing with and also make known their privacy policies. Meeting SOX and GLBA requirements enables the cloud service provider to be certified that they are compliant with these laws. This helps the potential cloud customer to meet their compliance requirements for SOX and GLBA while using the cloud service.

Table 5.3 Summary of US certifications for major cloud service providers

Service provider	HIPAA	SOX	GLBA	FISMA	SAS 70	PCI DSS
Amazon web services	Y	Y	Y	Y	Y	Y
Microsoft cloud	Y	Y	Y	Y	Y	Y
Google apps	Y	Y	–	Y	Y	–
Apple iCloud	N	–	–	–	Y	–
Rackspace	–	Y	–	–	Y	Y
Salesforce	Y	Y	–	Y	Y	–
Dropbox	N	–	–	–	N	N

– denotes unable to verify compliance or non-compliance

Cloud services are extensively used by governments at the federal, state and local levels. Since the US federal government use of cloud services is so extensive,
all major cloud service providers such as AWS, Microsoft, Google, Rackspace
and Salesforce have a separate government cloud that is isolated from their public
cloud offerings. All government business requires meeting the Federal Information
Security Management Act (FISMA) standards. In order for government agencies
as well as others providing service to the government sector to meet the FISMA
requirements the government cloud service provider should get FISMA certification. FISMA requires organizations to continuously monitor their assets, con-
figurations and vulnerabilities almost round the clock. It also requires strict identity management for all personnel having access to the systems. Since the cloud
service provider controls the infrastructure such requirements could be met only
if the cloud service provider is FISMA certified. Consequently, it is essential for
the government cloud service provider to have FISMA certification. Major cloud
service providers are committed to having their service comply with the required
government laws and industry standards so that their customers will be able to
meet their reporting requirements for compliance. We summarize in Table 5.3 all
the important certifications that major cloud service providers are complying with
now.

Cloud computing growth has been exponential over the past 5 years. The major
service providers all use proprietary technologies and this new technology tool does
not have a common global standard. This makes it difficult for cloud customers
to switch from one service provider to another for whatever reason. We saw in
the last section some effort to provide comparable data in a single source such as
the STAR registry, the Cloud Control Matrix or the Service Measurement Index.
Still, some form of third party audit of the cloud service provides the necessary
confidence building information for the cloud customers. In this regard most cloud
service providers seek the SAS 70 Type II Audit. This audit verifies the cloud service provider claims about internal controls, service uptime and security policies.
This type of service has been around for a long time. In order to meet the international reporting requirements this service has been revised and renamed as SSAE
16 (Statement on Standards for Attestation Engagements). It was launched in 2011
and cloud service providers who previously met the SAS 70 Type II Audit requirements are now embracing the SSAE 16 standard. This standard is only applicable

Table 5.4 Summary of international certifications for major cloud service providers

Service provider	SSAE 16	ISAE 3402	ISO 27001	Safe harbor	ISO 9001
Amazon web services	Y	Y	Y	Y	–
Microsoft cloud	Y	Y	Y	Y	–
Google apps	Y	Y	Y	Y	–
Apple iCloud	Y	–	–	Y	–
Rackspace	Y	Y	Y	Y	Y
Salesforce	Y	Y	Y	Y	–
Dropbox	Y	Y	Y	Y	N

– denotes unable to verify compliance or non-compliance

for the audit component of third party validation of service quality (SSAE16 2011). Associated with SSAE 16 is the international standard ISAE 3402, which gives the details on the global assurance standards for reporting on the controls at the service organization level (ISAE3402 2011). Most organizations that have the SAS 70 Type II Audit compliance are now updating their certification to the revised SSAE 16 and ISAE 3402 standards. Besides these standards, many service providers are complying with the ISO 27001, ISO 9001 and European Safe Harbor Agreements for data protection and privacy standards. These international standards enhance the security compliance aspects by focusing on associates set of standards. In Table 5.4 we summarize the status of global compliance certifications acquired by major service providers.

In addition to several government laws requiring security compliance, there are also other industry standards that have to be met in certain cases. The case in point is the Payment Card Industry Data Security Standard (PCI-DSS). This deals mostly with Point of Sale (POS) terminals having the adequate security safeguards. For those businesses such as car rental agencies and restaurants that use portable card swipe devices, additional mobile unit-to-base device communication encryption capability is required. If the cloud service provider meets the PCI-DSS requirements as part of their business processes then their customers will be able to meet their compliance requirements. For this reason the cloud service provider would need to acquire the PCI-DSS certification. Since payment cards are used globally and since the cloud computing service is used globally, it is essential for the cloud service provider to meet the PCI-DSS standards.

With the advancements in communication technology, many mobile devices are able to handle credit card data without a card swipe device. Customer is able to register their mobile devices ahead of time and run the special App that facilitates their use of credit card with mobile devices. The security associated with these Apps is now transferred to the mobile communications provider. The apps themselves must still comply with PCI-DSS. The POS merchants who have to bear the POS security cost and the associated audit, like the mobile apps payment mechanism because it saves them the cost associated with credit card processing. Since the payments are now handled through mobile devices without a card swipe, the future of mobile commerce (mcommerce) depends on the cloud's ability to handle secure transactions. There should be mechanisms to authenticate the user prior to the use of a credit card through a mobile device because such devices could be lost or stolen. The volume of mobile commerce is expected to grow from $ 42 billion in 2013 to $ 113 billion in 2017 (eMarketer 2013). This is a compelling reason for enhancing cloud security to support mcommerce applications.

5.4   Access Control

Cloud customers are fully cognizant of the fact that they do not have control over the cloud infrastructure. Consequently, they expect the cloud service provider to have suitable authentication tools to control access to the stored data. Moreover, the privileged users at the service provider management level should not be able to view or modify the customer data. Automatic logs of data access should be available to the customer in order to ease their concern that a rogue privileged user might have access to their data. The service provider should be transparent with the access control policies. By design the service provider manages the cloud infrastructure but does not own the customer data which resides in their servers.

The access control issue for cloud customers is applicable for the SPI model (SaaS, PaaS, IaaS) and the two major cloud deployment models—public cloud and private cloud. This means this problem is applicable to small, medium and large businesses that use cloud service. Multiple surveys have shown that over 75 % of the businesses worldwide use cloud services. Businesses that have a mature access control process in their internal systems have to rethink the way cloud access control is handled. Typically when a hybrid cloud approach is used by enterprises they use a private cloud for sensitive material and the public cloud for less sensitive applications. On the private cloud side the enterprise controls access and so it can use the existing internal control models. On the public cloud side the access control issues relate to how the applications are deployed. Because multi-tenancy and virtualization are quite prevalent in the public cloud the cloud customer is dependent on the cloud service provider controls for controlling access to their processes and data (Shackleford et al. 2013).

One common problem with access control is the type of authentication used to allow a user to have access to a device or service. Typical authentication mechanism used is user id and password combination. This is an older model that has been in use for over 50 years. With cloud technology there is greater granularity of access control needed. Typically when passwords are needed for several applications users tend to use the same password for all applications or write them down for later remembrance. Both approaches could introduce significant risks. One way to overcome this problem is by using Single Sign On (SSO). The SSO concept has given rise to an open standard called OpenID which enables the cloud user to have a single identity for multiple services. Before making all the cloud applications such as email, CRM, Calendar and Dropbox available to the user, using the corporate authentication procedure a one time password (OTP) is generated. McAfee provides such an application that could be run in the user's desktop that is integrated with the internal authentication process usually associated with Windows Active Directory. The OTP is the access control mechanism for all cloud applications. Once the user provides the OTP the cloud interface makes available all the applications that the user can access. Another way to control access to applications is using two-factor authentication. The validation data for two-factor authentication must be stored in internal systems.

Another situation where access control is needed involves users unknowingly releasing sensitive authentication data by using untrusted channels. This could lead to data loss through security breaches. Service providers should support data loss prevention (DLP) tools when customers attempt to store certain type of sensitive data in the cloud. For example, when an authorized user attempts to store a file with credit card numbers of customers from the internal secure system to an external file storage system such as Dropbox, the DLP tool would be able to prevent the file transfer. Typically encryption is used to store sensitive data in the cloud. This is not a good practice in light of the fact that using side-channel attacks hackers could retrieve the decryption key stored in the cloud (Ristenpart et al. 2009). In this case, using access control, sensitive information could be kept off of the cloud with suitable security policies. The enforcement of the security policies will prevent the storage of sensitive data in the cloud in the first place. Another important reason for having a cloud application providing access control to social media is to prevent the transfer of malicious malware onto the internal company network from the cloud. Businesses can mitigate such threat using threat intelligence and access control. Businesses should not use cloud service provider reputation as one metric in bypassing access control.

Closely associated with access control is Identity Management for access. OAuth is an open standard developed in 2007 for providing access control using open standards (OAuth 2007). Major cloud service providers such as Google and AWS have proprietary APIs that have similar functionality. Use of OAuth by the service provider will support customer mobility between cloud services. Identity and Access Management (IAM) allows the organization to control access to resources and the level of access to resources. For example, in large enterprises the organizational IT manages access control using a central repository such as the Active Directory. However, entities within the organization over a period of time start using SaaS services for tactical reasons. Such use bypasses the central access control repository. When privileged users from this group of SaaS users leave the organization, the central repository will not be aware of the elevated privileges the user had for specialized services. Using IAM on the other hand, organizations will be able to control access as needed. Federated Identity Management involves the use of existing identity verification systems and the Security Assertion Markup Language (SAML) helps with the management of the access control. SAML is an open standard used by many identity providers like McAfee.

Identity management in cloud computing is essential in order to provide secure and robust service using the cloud. He, Tran and Xie (2014) describe an architecture for identity management in Mobile Cloud Computing (MCC). A brief summary of this document is presented here to show the importance of IAM in MCC. The architecture consists of mobile network services (MNS) and mobile users (MUs). MUs connect to the cloud either through service provider APIs or embedded browser applications. The wireless link for MUs is either Wi-Fi or cellular. When a request emanates from an MU, the request and MU's location information are sent to MNS where Authentication, Authorization and Accounting (AAA) are performed. Since this takes place at the server end it has the necessary computing power and does not drain the battery life of the MU. Once the request is validated it is relayed to the service provider. During this process of AAA the system uses one of Location Based Access Control (LBAC), Role Based Access Control (RBAC) or Attribute Based Access Control (ABAC) relative to the user's request. This MCC architecture has the advantage of having access to large storage volumes of cloud storage once the access control is validated. Since the MU access is via the cloud, there is greater reliability in data access and the service provider has the ability to run anti-malware scanning on the data. The potential limitations of MCC are that the security and privacy concerns associated with the cloud service provider are inherited. Since the MU is resource-limited, a lightweight security framework has to be developed for running on the MU. Thus, IAM is feasible with mobile devices which dominate the access request origination points. Since businesses of all types allow end-user devices such as cell phones, mini Notebooks and Tablet PCs, a robust IAM process for mobile devices is essential in connection with the cloud (Takabi et al. 2010).

5.5   Organizational Control

In this section we will consider how organizations transition to cloud service and how their control systems change accordingly. As noted earlier cloud services are extremely popular with most of the small and medium sized businesses. Typically these are businesses with less than 250 employees. Most of the touted benefits of cloud computing are realistic for these businesses. However, large businesses such as major enterprises have the resources to manage their own computer system. Several of these large businesses are using cloud services for some applications but are still hesitant to trust cloud services because of loss of control and several well publicized outages and data breaches. For businesses of all types control of their operating environment is essential. We will analyze how businesses could retain some of the organizational control as they move their services to the cloud.

Cloud service uses the SPI model—SaaS, PaaS, and IaaS. The controls vary depending on the type of cloud service used. SaaS customers fully depend on the cloud service provider for all their computing needs. In order to realize the cost benefits an organization must choose the type of third party SaaS application. For example, small and medium sized businesses choose email, office productivity and database applications on the cloud. Microsoft and Google provide these types of services over the cloud. Small and medium sized businesses do not maintain an internal and external system. They transition their entire control mechanism to the cloud. For medium sized businesses the control system could be federated. For large

Table 5.5 Summary of organizational control for cloud services and onsite services

On site	SaaS	PaaS	IaaS	Legend
Data	Data	Data	Data	Business control
Apps	Apps	Apps	Apps	Dual control with CSP
VMs	VMs	VMs	VMs	CSP control
Storage	Storage	Storage	Storage	
Network	Network	Network	Network	

businesses the organizational control could involve a two-factor authentication. The cloud API for an application should be able to recognize the access credentials of a user. In order to protect the authentication data from getting transmitted through the cloud specialized APIs have been developed that support Single Sign On whereby a one-time password is generated based on the user credentials and the one-time password is what gets transmitted for organizational control to applications and data.

The question arises as to how the cloud services could enhance the trust of large businesses to move their internal computing operations to the cloud. The cloud service approach towards this end could be classified as follows:

1. Will the organization gain the most by moving services to the cloud?
2. Will the organization lose the least by moving services to the cloud?

In answer to item 1 the cloud service could point out the significant cost savings to the organization by moving capex to opex. Furthermore, the organization will be able to retain greater control over their services by using the Infrastructure as a Service feature of the cloud. When large organizations use the cloud for storage, they could use a hosted private cloud coupled with encryption of data at rest. Moreover, organizational control over security aspects is enhanced by using a Virtual Private Network (VPN) to access the cloud storage. The encryption and VPN aspects increase the cost of cloud service but in order to maintain the organizational control they are necessary. In answer to item 2 the cloud service could target small and medium sized businesses because they do not have the necessary internal resources to do what can be done cost effectively over the cloud. Since large businesses cannot entertain loss of control by moving to the cloud for cost savings, cloud services will have to target only the small and medium sized businesses. Since the number of small and medium sized businesses is large it would be economical for cloud services to work with them. We summarize the control aspects discussed so far in Table 5.5

The importance of organizational control arises when certain applications are moved from the organization to the cloud. In such instances if there is any database associated with the application, then it will be efficient to move the database also to the cloud and take steps to protect the database. One typical application that is often moved to the cloud is email. The authentication information for email server is located in IT directories that are synchronized with HR databases. When the email application is moved to the cloud, for faster response times the authentication information should also be moved to the cloud. For security purposes this database could be kept in read only format, thus providing faster lookup times. The cloud service provider should take precautions to protect this authentication database and prevent transfer of this file to any other application. Access control and data leak prevention will become top priorities for businesses when they move the services to the cloud. One solution available today is through the use of SAML and OAuth along with role based access control.

Finally, in the absence of global standards for cloud computing, a customer using a cloud service for certain applications should be able to revert the service from cloud to the organization. This happens when a cloud service provider goes out of business. Reverting to the business the application may not be a choice for many enterprises as they may not have the expertise in-house to run the application. In such cases a new set of organizational controls may have to be introduced when the reverted service is moved to a different cloud service provider because of proprietary APIs in the cloud.

5.6   Summary

Provisioning cloud computing security is the responsibility of the service provider. The service provider will be able to offer the necessary security to the applications and data in the cloud. The security aspects vary from service to service. SaaS applications are the most used service among the many public cloud offerings. Also, from a security perspective the customer lacks most control when using a SaaS application. On the other hand, customers have greater control when using the IaaS service. The customer should move only the low-risk applications to the cloud. Use of APIs facilitates accessing applications moved to the cloud. Using Single Sign On helps control access to the applications with suitable authentication. Using Data Loss Prevention techniques the service provider may have policies to evaluate all customer requests. Access control is essential to protect the applications as well as the stored data. Several new techniques such as OpenID and OAuth have been developed. Organizational control is essential when enterprise consider moving certain applications to the cloud gradually.

5.7   Review Questions

1. Describe the security issues associated with the SaaS, PaaS and IaaS service models.
2. Describe the cloud storage offerings of major cloud service providers.
3. Describe the way all major cloud service providers are distributing their data centers around the world for storage, backup and low latency of service.
4. Describe the security best practices that cloud service providers could adopt that would enhance cloud security.

5. What are the industry efforts to improve cloud security through various consortia efforts?
6. Describe the various cloud compliance requirements that the service providers could facilitate through their own certifications.
7. How could cloud service providers support access control requirements for security?
8. How could cloud service providers support organizational control requirements for security?

References

AWS. (2013). Amazon web services: Risk and compliance. Whitepaper. http://media.amazonweb-services.com/AWS_Risk_and_Compliance_Whitepaper.pdf. Accessed 22 Jan 2014.

Blum, D. (2009). Cloud computing security in the enterprise. Gartner research report.

Carpenter, M., Liston, T., & Skoudis, E. (2007). Hiding virtualization from attackers and malware. *IEEE Security and Privacy, 5*(3), 62–65.

CDW. (2012). Energy efficient IT report. http://www.cdwnewsroom.com/2012-energy-efficient-it-report/. Accessed 22 Jan 2014.

Chow, R., Gotlle, P., Jakobson, E., Staddon, J., Masuoka, R., & Molina, J. (2009). Controlling data in the cloud: Outsourcing computation without outsourcing control. Proceedings of the 2009 cloud computing workshop on cloud computing security.

Cloud Commons. (2011). http://www.ca.com/us/news/press-releases/na/2011/ca-technologies-unveils-cloud-commons-marketplace-and-developer-studio.aspx. Accessed 1 OCT 2014.

Cloud Security Alliance. (2013). Cloud controls matrix. https://cloudsecurityalliance.org/download/cloud-controls-matrix-v3/. Accessed 22 Jan 2014.

COSO. (2012). Enterprise risk management for cloud computing. http://www.coso.org/documents/Cloud%20Computing%20Thought%20Paper.pdf. Accessed 22 Jan 2014.

CSA STAR. (2011). Security, trust and assurance registry. https://cloudsecurityalliance.org/star/. Accessed 22 Jan 2014.

eMarketer. (2013). US retail Mcommerce sales 2011–2017. http://www.emarketer.com/. Accessed 1 Oct 2014

European Union Report. (2009). Cloud computing: Benefits, risks and recommendations for information security. http://www.enisa.europa.eu/activities/risk-management/files/deliverables/cloud-computing-risk-assessment. Accessed 22 Jan 2014.

Gartner. (2009). Data center efficiency and capacity: A metric to calculate Both. Gartner research report.

Hashizume, K., Rosado, D., Fernandez-Medina, E., & Fernandez, E. (2013). An analysis of security issues for cloud computing. *Jl. of Internet Services and Applications, 4*(5), 1–13.

He, B., Tran, T., & Xie, B. (2014). Authentication and identity management for secure cloud businesses and services, Chap. 11 in the book Security, Trust, and Regulatory Aspects of Cloud Computing in Business Environments, Editor S. Srinivasan, Hershey, PA: IGI Global.

ISAE3402. (2011). ISAE 3402 Global Standard. http://isae3402.com/. Accessed 22 Jan 2014.

McKinsey. (2008). Revolutionizing data center energy. McKinsey company report.

NIST. (2011). *Guidelines on security and privacy in public cloud computing, SP 800-144*. Gaithersburg: NIST Publication.

NIST. (2013). National vulnerability database. http://nvd.nist.gov/. Accessed 7 Jan 2014.

OAuth. (2007). OAuth standard. http://oauth.net/about/. Accessed 22 Jan 2014

Ponemon Institute. (2011). Cloud security study. http://www.dome9.com/resources/ponemon-cloud-security-study. Accessed 5 Feb 2014.

Ristenpart, T., Tromer, E., Schacham, H., & Savage, S. (2009). Hey, you, get off of my cloud: exploring information leakage in third party compute clouds. *Proceedings of the 16th ACM computer and communications security*, 199–212.

Rittinghouse, J., & Ransome, J. (2009). Security in the cloud: Cloud Computing Implementation, Management and Security. Boca Raton, FL: CRC Press.

Sengupta, S., Kaulgud, V., & Sharma, V. (2011). Cloud computing security—trends and research directions. IEEE world congress on services, pp. 524–531.

Shackleford, D. (2013). Simplifying cloud access without sacrificing corporate control. SANS Whitepaper.

Softlayer. (2013). Bare metal servers. http://www.softlayer.com. Accessed 22 Jan 2014.

Srinivasan, S. (2014a). Is security realistic in cloud computing? *Journal of International Technology and Information Management, 13*(1).

Srinivasan, S. (2014b). *Security, trust, and regulatory aspects of cloud computing in business environments, Chapter 8*. Hershey: IGI Global.

SSAE16. (2011). SSAE 16 standard overview. http://ssae16.com/SSAE16_overview.html. Accessed 22 Jan 2014.

Takabi, H., Joshi, J., & Ahn, G. (2010). Security and privacy challenges in cloud computing environments. *IEEE Security & Privacy, 8*(6), 24–31.

TIA. (2012). Data center standard. http://tiaonline.org/node/773. Accessed 22 Jan 2014.

Zhang, Y., Juels, A., Reiter, M., & Ristenpart, T. (2012). Cross-VM side channels and their use to extract private keys. *Proceedings of the 2012 ACM conference on computer and communications security*, 305–316.

Chapter 6
Assessing Cloud Computing for Business Use

Abstract Cloud computing is widely accepted by businesses of all sizes. However, the level at which the adoption of cloud computing has moved varies significantly based on company size. Small businesses that lack the people or financial resources to manage a computing system mostly have embraced cloud computing without too much reservation. Medium sized businesses also benefit from transferring the management aspects of a computing system to the cloud. Large businesses have an option to use their internal systems exclusively or to transition some services to the cloud and keep the more sensitive applications in-house. This hybrid approach is expected to last for a while. Businesses consider cost savings as an attractive feature but require more control over their infrastructure. For this reason the private cloud deployments are growing rapidly. Cost comparison for cloud service versus in-house computing environment shows that private clouds are still cost effective. In this chapter we assess the reasons for companies moving to the cloud. This assessment includes cost, availability of service, reliability and security. We study in detail the cost factors of cloud computing and how it could help with cost savings for businesses. We analyze the risk components of cloud service and discuss how businesses could mitigate the cloud risk by using third party controls. One of the important things to note is that businesses tend to use a form of cloud service indirectly. We analyze in depth how outsourcing compares with cloud computing.

Keywords Cloud computing · Benefits · Risks · Cost factors · Outsourcing · Cloud storage · Data breach

6.1 Introduction

Cloud computing's benefits are not disputed. Businesses realize that they have at their disposal all the computing services that they need and they will pay only for what they use. This is ideal for cost containment for a business. For many businesses the focus is on their core strengths. At the same time the businesses realize the importance of having a good computing system with the latest hardware and software technologies. If the business were to develop this capability on their own they would not only have to spend a large sum of money but also take their focus away somewhat from their core strengths. For many small and medium sized businesses allocating substantial financial resources to computing would be difficult. For this

S. Srinivasan, *Cloud Computing Basics,* SpringerBriefs in Electrical and
Computer Engineering, DOI 10.1007/978-1-4614-7699-3_6,
© Springer Science+Business Media New York 2014

reason, when cloud computing came along as an option, they opted for the cloud service overwhelmingly. For small businesses the tradeoff is that they get access to high end computing technology at a fraction of the cost. Moreover, since the technology will change rapidly, they do not have to keep up with updating and managing the technology. Small businesses are aware of their data being on the cloud and the associated risks. According to the National Small Business Association 2013 survey, 43 % of small businesses use cloud computing and this is a significant jump from the 2010 cloud usage rate of 5 % (NSBA 2013).

Most of the small businesses are in manufacturing certain parts for a larger company or into services. Small businesses with fewer than 50 employees tend to use information systems primarily for email and record keeping. Since their needs are very minimal they do not try to use the cloud service. Small businesses with more than 50 employees use cloud services more. Over 6 million small businesses used the cloud service as a way to make their presence known. The analogy for small businesses in the use of cloud IT is similar to using package delivery services such as UPS and FedEx to augment their core strengths. Since small businesses usually partner with a large company and provide a single product, they focus more on keeping the cost down. When large manufacturers in the automotive sector want their suppliers to use certain advanced technology, the only cost effective solution for the small companies is the use of cloud services. With supply chain becoming critical for large businesses, the small business partners tend to use the cloud services in order to have access to advanced technologies.

In order to use cloud services a business must be aware of what is available on the cloud and how it would benefit the business. For a working definition, we would classify a business with less than 250 employees as Small Business. Businesses with 250–500 employees as Medium Sized Business and with over 500 employees will be classified as a Large Business. Enterprise level businesses tend to have several thousand employees spread over multiple countries. Of the three types of standard services available on the cloud—SaaS, PaaS, and IaaS—only SaaS is most appropriate for small businesses. Many businesses that use PaaS are application developers. Large businesses tend to use IaaS more than the other two types of services, even though they use specialized applications of SaaS such as Customer Relations Management (CRM) over the cloud and PaaS service for development testing. There are two primary deployment models that are popular with many businesses. These are public cloud, the most used by small businesses, and the private cloud, most used by large businesses. Cloud service gives the opportunity for many entrepreneurs to be service marketers on the cloud without having to manage any cloud of their own. Two other deployment models that are also prevalent are the hybrid cloud and the community cloud. The hybrid cloud helps primarily large businesses as they use the private cloud located on premise for sensitive applications and use the public cloud for less sensitive applications to meet their service needs. The community cloud is focused on industry verticals such as health care, automotive and finance sectors. Third party service providers focus their expertise in a particular industry such as health care and make the services comply with requirements such as HIPAA. By keeping their service open only to a specific industry segment they are able to enhance data security and service sharing. One unintended benefit of the community cloud is akin to the growth of internet in the early years when only certain major organizations belonged to the internet and so there was a shared security.

The largest of the cloud service providers is Amazon Web Services (AWS). Many entrepreneurs contract for AWS services in order to provide a focused and beneficial service to small businesses that lack the expertise to know what they would need from a cloud service. If a small business needs an enterprise level application then the place to get that application is on the cloud because it would be prohibitively expensive otherwise. Medium sized businesses tend to use the cloud both for SaaS and PaaS. However, public cloud is still their choice for services because of the many benefits it offers. Unlike small businesses that have minimal security concerns when using a cloud service, medium sized businesses take steps to protect their data on the cloud. Medium sized businesses also work with large businesses as partners, especially in manufacturing. The three industries in which medium sized businesses have a significant presence are Health Care, Financial Service and Manufacturing.

Large businesses and Enterprise level organizations use all three types of cloud services—SaaS, PaaS and IaaS. They have the expertise and resources to manage their computing platforms. However, these large businesses prefer to have capital fluidity and so using cloud computing gives them the ability to switch the Capex costs to Opex costs. They tend to use private clouds more than public clouds because of their sensitivity to data security and data loss prevention. Many large businesses tend to use public cloud services for less sensitive applications and customer facing services. Their primary need for PaaS service is for testing purposes using a variety of operating environments. Because large businesses can handle their own computing systems and because of security concerns with cloud data storage, the cloud adoption rate among large businesses is relatively lower when compared to small and medium sized businesses. Gartner Research and other research studies show that major cloud service providers are not focused on large businesses for promoting the cloud service.

Cloud computing is maturing slowly, with new standards being developed that will facilitate service switching. Even though theoretically it is feasible to switch service from one cloud service provider to another, it has not been practical. One typical issue is the format in which data is stored in cloud service. In the absence of global standards, major service providers such as AWS, Microsoft and Google use proprietary APIs. There is a significant push towards open standards in the form of Open Stack Operating System and business consortium efforts such as the Cloud Security Alliance (CSA) guidelines (Open Stack 2011; CSA 2014). The Open Stack cloud community spans over 130 countries and is supported by major cloud service providers such as Rackspace and HP.

6.2   Benefits to Business

Acceptance of any new technology hinges on the benefits to the organization. Cloud computing offerings started in 2006. Since then the service is promoted as shifting the management of an information system from the organization to the cloud service provider. Inherent benefits of such a move is the service availability, meeting demand elasticity, metered service, security and access to advanced services. All these claims are real benefits to an organization and so the cloud adoption rate has steadily increased over the years to over 75% with a Compound Annual Growth Rate (CAGR) of 25%. In certain developing countries such as India the CAGR is projected to be close to 35% (Gartner 2013) The cloud adoption is not limited to U.S. only. In Europe, the cloud adoption is significant both with businesses and governments (Interxion 2013). Virtualization is at the top of cloud services. This concept has helped with the significant increase in server utilization rate. As mentioned earlier, the adoption of a particular type of cloud service varies by size of the organization. Small businesses tend to use Software as a Service more. Medium sized businesses use Software as a Service and Platform as a Service more. Large businesses tend to use Infrastructure as a Service more. Public cloud deployments are popular with small and medium sized businesses where as large businesses tend to favor private cloud deployments because of the service control and security it offers. With the growth in technology the trend is that hybrid clouds will dominate the adoption patterns by large businesses. Gartner Research projects that by 2016 hybrid cloud will be the principal cloud deployment model for large businesses. It is to be noted that large businesses serve two roles—cloud users and cloud providers. In their role as cloud users they focus on how, when and where to use the cloud service but not on the management of the hardware and software that make up the cloud. In their role as cloud providers they focus exclusively on the management aspects of the hardware and software.

Businesses of all sizes benefit from the cloud service. Small and medium sized businesses that could not afford high end computing services are able to use such services over the cloud. Because of economies of scale the cloud service providers are able to provide such services on a pay-per-use basis. Major businesses such as Microsoft and Adobe have started offering their primary products over the cloud without the complex licensing requirements. Microsoft Office 365 consists of all the applications in their Office Suite and it is available over the internet. Consequently the end user need not keep up with the upgrades to the software as well as service patches that come regularly. Adobe's Creative Cloud provides access to their popular PDF and Photo Shop suite of products. It facilitates collaboration with other users and sharing the creative works easily with thousands of other users over the cloud quickly. In a non-cloud setting these would be prohibitively expensive for the ordinary user but also will not have the ability to collaborate and share their work with others.

Cloud computing's primary strategy of pay-as-you-go is both a benefit and drawback. As this industry matures it will focus on the need to have price stability for businesses. It is true that businesses save cost using the pay-as-you-go model because they pay only for what they consume. However, what is not quite apparent is the cost of the service. A business that starts out using a particular application service such as CRM does not have the price protection guarantee over an extended period of time that is common in contract services. Furthermore, even with SLAs the cloud service providers change the terms of the contract with 30-day or 45-day notice. In the absence of the flexibility to move their service to a different service provider customers are forced to accept the higher cost for the same service. This hidden aspect of cost spike over a short period of time makes the cloud customer not realize the cost savings anticipated in using the cloud service. Studies have shown that the cost savings of Software as a Service disappear after 2 years. The cloud computing contracts generally favor the service provider.

Core requirement for any business is the ability to communicate with their internal employees and customers quickly and easily. Like other popular email products such as Gmail and Yahoo Mail, Microsoft Exchange offers the Outlook mail server. With over 400 million users worldwide, Microsoft Exchange is used by businesses of all types. Managing an Exchange server is both complex and expensive. Since this service is available over the cloud, businesses can subscribe to the service easily and pay for what they use. Likewise, another popular application is database services. SQL Server is one such database system and it is available over the cloud. Another important cloud service has been the Customer Relations Management (CRM) application. Salesforce is the leader in this multi-billion dollar CRM application service. These trends clearly show that more and more popular services are offered over the cloud and so businesses have the ability to subscribe these high end services at a fraction of the cost (Catteddu, 2010).

Typically medium sized and large businesses are very concerned about security of cloud service. The concern is more widespread among large businesses more than the medium sized businesses. Usually small businesses do not attach great importance to security aspects of the cloud service. Surveys show that the cloud service adoption rate is lower among large businesses because of their security concerns. These concerns are addressed by cloud service providers by making available suitable APIs that support security. The primary requirement for large businesses is their ability to control access to their stored data in the cloud. The newer APIs extend the organizational access controls through the use of Single Sign On whereby a temporary one time password is generated based on the customer identity. Such one-time passwords protect the organizational control over user identity and they are not shared over the cloud. One such API is McAfee Pledge Software Token (McAfee 2014).

Another major benefit of cloud service is the high level of automation built into service provisioning. This helps the cloud service customer to use on demand computing resources and storage capacity. From the user perspective the automation is a significant benefit since the customer can configure their cloud service rapidly and deploy them quickly. A classic instance of rapid deployment of additional resources as needed is highlighted by the Animoto experience. As a cloud service provider for a specific application, namely a video service, Animoto experienced tremendous growth. It needed to grow from 50 servers to 3,500 servers over a 3 day period using AWS' Elastic Computer Cloud (EC2) service. Since its AWS cloud service had high level of automation it was able to use the necessary computing power quickly and meet its demand (Bezos 2008).

The data growth has been exponential for many businesses. Even though the cost of storage has come down dramatically, costing only a few cents per gigabyte of storage, the associated cost of protecting and managing stored data is significant. Typical with any data storage is backup and recovery as well as security of data at rest. These are all expensive propositions for a business to undertake. A cost effective alternative is the cloud storage. Amazon Web Services provides a very popular storage service known as the Simple Storage Service (S3). Other storage services available in the cloud are Microsoft's SkyDrive, Google Drive, Dropbox, Apple iCloud and Box. All these services offer some free storage and then a tiered service plan. The primary benefit of using cloud storage is not only the benefit of expanding storage when needed but also contracting storage when not needed. The cloud storage service provider is responsible for data backup and recovery. The security of data storage is the responsibility of the cloud service provider with certain exclusions. It is important to realize that the cloud service provider will not be aware of the sensitivity nature of the data stored in their devices. It is up to the businesses to restrict access to the storage through access control. Typical access control methods available are Role Based Access Control (RBAC) and Attribute Based Access Control (ABAC). RBAC helps the business with a simplified access management. The ABAC model, on the other hand, is a logical access control model that evaluates the access request against the attributes of entities, objects and the environment before granting or denying a request (NIST 2013). ABAC is more fine-grained than RBAC. The service provider's obligations for data protection are limited to unauthorized access of stored data because of multi-tenancy.

Cloud services support business continuity aspects by storing customer data in a location different from where the business is located. Since the business accesses their cloud service using the Internet, they could be up and running from a different location after a disaster because the access methods did not change. Thus, cloud services inherently provide business continuity feature automatically. However, this should not be taken for granted as more businesses are dependent on cloud storage and they sometimes become unavailable due to natural disasters such as the Japanese tsunami and the effect of hurricane Sandy on the US East Coast with regard to utility power. Based on the possibility of disasters of this scale some companies have developed more resilient processes to guarantee business continuity. One such is Netflix's Chaos Monkey process which helps the company randomly remove one key component of the service plane and still expect the service to continue operating. This proactive step helped Netflix continue its service without interruption when AWS experienced a significant service outage in one of its US East Coast data centers which Netflix uses for its cloud offering. It is important to recognize that AWS provides service to other major cloud service providers such as Dropbox, Foursquare, Pinterest and Flickr.

Another benefit for businesses in using cloud services is their ability to comply with government regulations and industry standards. This is because the cloud service providers acquire compliance certifications for HIPAA, SOX, GLBA, FISMA, SAS 70 and PCI DSS. When a cloud customer needs to provide any of these compliance data, they are able to access the necessary data logs and gather the required data to satisfy their compliance requirements. In some cases, the cloud service providers also acquire international certifications such as ISO 27001, Safe Harbor Agreement Compliance, and ISAE 3402 Audit Standard compliance. These international certifications help the large businesses because of their global presence. We conclude this section with a summary of the business benefits discussed above, based on the type of business (Table 6.1).

Table 6.1 Summary of business benefits for use of cloud computing

Business benefit	Small business	Medium sized business	Large business
Service availability	Y	Y	Y
Service reliability	Y	Y	Y
Meeting demand elasticity	Y	Y	Y
Ability to pay-as-you-go	Y	Y	Y
Service automation	–	Y	Y
Email support	Y	Y	–
Database support	Y	Y	–
Customer relations management support	–	Y	Y
Access control support	–	–	Y
Security	–	Y	Y
Business continuity	–	Y	Y
Data storage	–	Y	Y
Data backup and recovery	–	–	Y
Meeting regulatory compliance	–	–	Y
Meeting industry compliance	Y	Y	Y

–denotes that the benefit is not significant

6.3   Risks of Cloud Computing

Cloud computing helps the business growth seamlessly through easy service provisioning. At the same time, the cloud service introduces several types of risks that the business must be prepared for when using the cloud service. Some of these risks are unique to cloud service as they are not part of a customer controlled computing environment (Srinivasan 2014). The *first risk* is the lack of control over the computing infrastructure. This includes the possibility of data leakage during processing. This arises because of server virtualization which supports multi-tenancy of different customers on a single physical server. Virtualization is at the heart of cloud services and the service providers take precautions to protect data leakage. But, businesses have to treat this risk seriously. Small and medium sized businesses may not have an alternative for this. Large businesses may be able to overcome this risk through the use of a private cloud. A *second risk* is due to security and privacy controls on the cloud. Cloud service providers do not consider that it is their responsibility to protect and secure customer data. This is one of the conclusions of the Poenmon Institute survey on cloud security (Ponemon Institute 2011). The primary reason for this thinking on the part of the cloud service provider is that they are unaware of the nature of data stored in their infrastructure. Since it is a security and privacy risk, the cloud customer is responsible for protecting this information (Weinman 2012, Gellman 2009). Encryption is one method available for this type of protection. In that case the encryption key should be stored at the customer site.

A *third risk* involves service management by the cloud service provider. This relates to the number of privileged users at the service provider who have access to the servers in which the customer's applications and data reside. Also, the service provider should create automatic logs of all service provider access to server instances. Next, the cloud customer should be able to access this data on demand. Even though the customer may not be able to control this risk, at least the customer will be able to point out the cause for any data loss should a rogue administrator cause such data loss. The *fourth risk* relates to regulatory compliance. This risk should not be of major concern for the cloud customer because the cloud service providers invariably seek such compliance certifications on their own because of competition among the providers and maintaining their reputation. The typical compliance certifications come from Health Insurance and Portability and Accountability Act (HIPAA), Sarbanes–Oxley Act (SOX), Gramm–Leach–Bliley Act (GLBA), Federal Information Security Management Act (FISMA), SAS 70 (Statement on Accounting Standards 70) Type II Audit, PCI-DSS (Payment Card Industry—Data Security Standard) requirements, and Safe Harbor Agreement with the European Union concerning data privacy protection. However, the cloud customer who has a need to provide evidence of compliance for any of these requirements should be able to gather the necessary data from the service provider through automatic logs and security policies in force. With many major and minor cloud service providers offering their cloud service globally, they may need to have some global compliance certifications as well. Some such certifications available are ISAE 3402 and ISO 27001. ISAE 3402 is simply the international equivalent of customer controls that is measured by the SAS 70 Type II Audit extension, known as SSAE 16 (Statement on Standards for Attestation Engagements 16).

A *fifth risk* is cloud outages and service availability. This risk is no less when the data is kept in-house by companies. The question is how quickly the company can access the data stored for its business use. Infrastructure capacity and resource availability favors the cloud service provider to restore service faster than the cloud customer if they were to host the data in their own data centers. A recent study by a panel of industry and academic experts on outages concluded that the major outages since 2007 among 13 cloud service providers resulted in a total downtime of 568 h with an economic loss estimated at $ 72 million (Kopytoff 2012). Most cloud service providers tout their service availability at 99.95 %. This would mean that the service could only be down 21 min per month. Reality is that the service providers are not able to offer such high service availability. The consequence is that they have to credit the customer for missing the service guarantee. Cloud service providers have modified their guarantees in one of two ways to deal with this problem. Either they require the customer to choose multiple Availability Zones for the SLA (means higher cost to the customer) or change the measurement period for 99.95 % from month to a year. This simple change has allowed the service provider to have up to 4 h downtime per year, thus averaging out their service availability over 12 months instead of per month. Such outages cause serious reputation issues for the business, as the end user would hold the application provider responsible and not be worried about where the application is emanating from in order to provide the service. Related to this risk is data breach, which is the *sixth risk* that we consider. Over the past 5 years data breaches have occurred with or without the use of a cloud service provider. When the data breach occurs at the cloud service provider level it involves many businesses, not just one business. Every one of these incidents affects the reputation of the cloud customer and possibly leads to identity theft of its end users. For this reason the data breaches also have a financial consequence for the cloud customer. These two risks discussed above related to the availability of customer data when needed. Along the same lines there is a much more serious risk where the cloud service provider is unable to provide the service due to financial reasons or violation of some law that resulted in the service being shut down by law enforcement. We treat this as the *seventh risk* facing the customer. One such incident happened when the cloud storage company Nirvanix shut down in 2013 and all its customers had to get their data out of Nirvanix in a hurry.

Cloud services promote their service as pay-as-you-go. Theoretically this gives the cloud customer the freedom to use any cloud service they like. In reality, most cloud services require a certain period of commitment for an application such as CRM or email. Moreover, during the time the customer uses the service the associated data gets stored in proprietary format of the service provider. Even if the customer were to think of migrating to a different cloud service provider, moving the existing data in a readable format is time consuming and expensive. The one possible exception to this is the use of AWS cloud service. Since AWS is the largest cloud service provider, many other cloud service providers provide APIs for transfer of service from AWS, should the customer want. The above discussion is the *eighth risk* that we consider and it is commonly known as data lock-in (US-CERT 2012).

Cloud service providers, because of their many compliance certifications and voluntary reporting to the Cloud Security Alliance's Security, Trust and Assurance Registry (STAR), have several log data. The *ninth risk* is related to lack of access to this kind of data. Cloud service providers are reluctant for third party audit of their infrastructure management and policy enforcement. Since they serve many customers it would be a major distraction if they were to subject themselves to every customer's third party audit. However, they should be willing to share all such log data that they have that would satisfy the customers' needs. Since this is not a standard practice the cloud customer should insist on such data as part of their Service Level Agreement (SLA) prior to signing up for cloud service.

We will conclude this section with a brief analysis of the ways to mitigate the risks identified above. One way to mitigate the risk of service becoming unavailable when needed is to use a second cloud service provider for data backup, not the same provider. This approach also helps with possible switching of service later to this service provider. The credits offered per SLA violation is miniscule compared to the loss of income to the business when the site is unavailable. It would be better for cloud customers to have independent insurance against loss of income due to service outage. Cloud service providers are cooperating with insurers so that they could examine the cloud service operation and write their insurance policies accordingly. Another risk to mitigate is security. Since security concerns are important to address, the customer should explore the possibility of having a Virtual Private Network (VPN) access to the service provider with strong encryption for data at rest. This adds cost to the service but at least provides the extra security protection. Another option is to consider a Virtual Private Cloud (VPC). Service providers like AWS provide VPCs at an affordable cost. There are also Cloud Service Brokers (CSBs) who lookout for the interests of the customer and identify service providers who are reliable and have solid security practices. Moreover, the CSBs can help steer several customers towards an affordable service by negotiating an affordable service cost based on overall volume. CSBs are popular in the cloud service industry as they account for over $ 500 million of service cost for the customers. The benefits derived from the use of CSBs far outweigh the cost of their service. Many cloud service providers use third party service contractors for specific services. In such cases customer data is transferred to the third party contractor without notification to the cloud customer. Even though the cloud service provider is responsible for data protection, data breach potential increases with the exposure to more providers involved in the service loop.

6.4   Cost Factors in Cloud Computing

One of the key benefits of cloud computing for businesses is the cost factor. Small and medium sized businesses especially place higher emphasis on the cost savings aspect of cloud computing because they are able to get high-end computing at an affordable cost. The cost component could be divided into explicit costs for service and the implicit costs associated with the service. Explicit costs are deterministic and the implicit costs have to be estimated. For example, selecting a certain amount of storage for a given period of time has a fixed cost whereas the steps a company has to take to secure their cloud data is an implicit cost. One reason for this is that the resources allocated for such a purpose is likely to be usable in other settings as well and so the cost has to be distributed to other areas where the resources are used. A 2013 study commissioned by Research in Action shows that 79 % of CIOs are concerned about the hidden costs of managing applications in the cloud (Research in Action 2013). In this section we will review both the explicit and implicit costs associated with cloud computing service.

Cloud service providers promote their service as pay-as-you-go which implies that no contract is required. In reality it is not the case. The "pay-as-you-go" model comes with a minimum service usage requirement. For example, the largest of the cloud services is Amazon Web Services (AWS). The AWS contract requires a customer signing up for a public cloud service with certain applications through their Software as a Service to use the service for at least 30 days and use two different Availability Zones in order for the service availability guarantee to be enforced. To understand better the cloud service costs let us introduce certain terms that we would refer for service cost. An Availability Zone (AZ) is like a data center. A region consists of several AZs. Within an AZ there could be compute blocks that are allocated for cloud services and cost is measured based on the number of compute blocks used. In a public cloud scenario the customer pays for compute blocks used within an Availability Zone. Some of the major service providers such as Rackspace and Microsoft reserve the right to change the contract with a 45 day notice. When customers do not have the ease of switching the 45-day notice does not mean much, other than a cost increase. Moreover, by requiring two Availability Zones in

Table 6.2 Summary of cloud storage costs

Service provider	Free storage limit	Additional storage cost/year	Remarks
AWS	5 GB	1 GB—$ 1.14	15 GB of data transfer out is free
Google drive	15 GB	100 GB—$ 60 200 GB—$ 100 400 GB—$ 240	Pure storage service without additional synchronization and sharing features of Dropbox
Microsoft skydrive	7 GB	50 GB—$ 25 100 GB—$ 50	Pure storage service
Rackspace	None	1 GB—$ 1.44 Std 1 GB—$ 6.00 SSD	Std volume uses SATA technology. Solid State Device (SSD) storage uses flash memory and is faster
Salesforce	None	1 GB—$ 1,500 data	Data storage deals with records. File storage deals with files
		1 GB—$ 60 file	Each record in CRM takes 2 K bytes

order for the SLA to be enforced for availability, the service provider is expecting the customer to pay for more than what they need in the form of a second Availability Zone. AWS, Google and Salesforce also have similar requirements, but with a 30-day advance notice to change the contract. These five companies account for an overwhelming majority of cloud services and so we reviewed their costs. It is important to note that, except for Salesforce, none of the other companies put a cap on the cost increase. That makes it difficult for customers to anticipate the cost of cloud service over a longer period of time.

Another teaser from these companies related to cost is based on a limited free service tier. Considering the storage cost on the cloud, all major cloud service providers offer a free storage tier. Table 6.2 shows a comparative view of these costs. First, all companies will need far more storage space than that offered though the free storage tier. Second, in the absence of global standards, the storage API is proprietary. Cloud service providers charge for data egress always and in some cases charge for data ingress as well. These are hidden costs. Third, data storage is not for a short period of time. For all practical purposes the storage space used is indefinite and growing, even if certain data get deleted after a period of time. Thus, the storage cost alone will become significant over a span of 5 years. Moreover, if the cloud customer decides to move the storage to another storage service provider then the cloud customer has to pay for the bandwidth needed to move the data. Assuming a petabyte of stored data (1 million GB) it would take over 3 months to transfer the data at the data transfer rate of 1 Gbps. Since this would not be acceptable for many businesses a higher data transfer rate would be needed and that would increase the cost of moving data from one storage provider to another.

Cloud service provider in their IaaS service provides the customer with the hardware. It is up to the customer to run the applications of their choice in that hardware.

Many customers find the cost of application integration is significant. A large business would want integration of their email system with their voice mail system. One benefit of this integration is that voice mails could be delivered as email and vice versa. Also, when a voice mail is replicated by an email then the voice mail could be deleted automatically. To address the integration cost issue in the cloud SAP has developed HANA (High-performance ANalytic Appliance) that combines database and application capabilities in-memory for a high-speed response. Another hidden cost of cloud computing involves the need to test applications for cloud use. If the application provider does not provide this feature then businesses that migrate their services to the cloud will have to develop the necessary interfaces. Two other hidden costs are with respect to set up of cloud service and service trials. The cloud service provider emphasizes the ongoing cost advantages of the cloud service but the customer has to transition their service to the cloud and there is significant cost involved with that. Often service trials turn into service if the business does not watch out for the conditions under which a trial becomes a service. Businesses have to realize that once they move a service to the cloud the cost savings up front may not last long. This is because the service will be in use for many years and so the cloud costs will be recurring. Some businesses are finding that using cloud services for the longer run is not cost effective.

Cost of IT services is closely watched by businesses. That is one reason cloud computing is closely looked at by businesses, especially small and medium sized businesses. In this section we have concentrated so far on how much services cost in cloud computing. The ultimate goal is to have cost savings. With this in mind we can look at the evolution of cloud computing at an industrialized service scale. Businesses saved up to 30 % by outsourcing and offshoring IT services. Because of high levels of customization and people centered activity, the savings could not increase or the service scale up. Cloud computing on the other hand was designed with standardized service offered over the internet with high levels of automation. This provided comparable savings but had the additional advantage of scaling up. The cloud computing industry is working towards global standards that will help offer industrialized services that help the customer to switch between service providers easily (Cloud Industry Forum 2014). The two areas where this savings would be noticeable are email service and storage over the cloud. The barrier to significant growth in these two areas is due to the security and privacy concerns. Major cloud service providers such as Amazon, Microsoft and Google are focused on consumer rather than corporate clients. Cloud computing will emerge over the next 5 years with many security and privacy compliance features built-in. This will accelerate the adoption of more cloud services leading to significant cost savings for businesses.

So far we looked at several of the explicit costs associated with cloud computing. There are some implicit costs that also impact the continued use of cloud service. One such involves security. Businesses have a robust authentication mechanism for access control for their employees based on in-house management of this service. It will not be advisable to move this service to the cloud for security reasons. Moreover, businesses want to safeguard their authentication data. One way to accomplish this is to use a third party API which integrates with the internal authentication system and generates a one-time password that is used to control access to other applications on the cloud. If VPN is used for secure cloud access, then that cost should be distributed to other applications that also use the VPN service. Site mirroring is another service that benefits cloud service availability but also helps with email service availability and the cost benefit extends beyond the cloud service. Typical variations in service agreements are with respect to public versus private clouds. In public clouds the risk factor corresponds to the grace period at the end of the contract for accessing the data. Typically 30 days are common during which time the data move takes place and during which time the data will be accessible for the customer for their applications. There will be cost for such access. The more serious risk is with regard to data integrity and data confidentiality, which the cloud service provider does not want to assume. Thus the cloud customer will incur additional cost to provide these two critical assurances for their data.

6.5   Variation on Outsourcing

Cloud computing technology is viewed for its widespread availability, reliability and cost savings. In using the cloud service the business cedes control to the cloud service provider. Traditionally businesses let a third party provide a service based on their expertise for a fee. For example, pay roll is an outsourced service with many small businesses. In a similar way we can consider the cloud service as an outsourcer for the business. In this section we will review how cloud computing could be viewed as an outsourced service.

There are several similarities and differences between outsourcing and cloud computing. Traditional outsourcing takes over a prior function that was performed in-house and manages it. The outsourced function may not be done any differently by the outsourcer. The customer cedes control on that function to a third party who gets the job done. From the business perspective it is done by someone who has an expertise in that area. For example, a business that used to process customer credit cards for charges to their products or services outsources the credit card processing to a third party that specializes in that area. The outsourcer usually brings in more expertise and resources to the function. In a similar way the cloud service provider brings more infrastructure resources through economies of scale and offers the cloud service to the cloud customer. In the traditional outsourcing model the customer retains control over the outcome of the outsourced service. They have the ability to intervene much faster. However, in the cloud computing model the customer lacks control over the computing infrastructure. Even though theoretically they can move the service in-house, it is not practical. The cost factor in that case would be very high for the customer and they will lose valuable time providing the service. In other words, service lock-in is not very common with outsourcing whereas it is much more prevalent with cloud services.

One other similarity between outsourcing and cloud service is that they both can offer their specialized service to a variety of customers. This capability provides them with the ability to have economies of scale advantage in pricing their service. For example, the payroll giant ADP provides payroll services to a large number of customers globally. Because of their size they are able to command a better price from their suppliers. Both the outsourcer and the cloud service provider have the ability to provide data to the customer to meet their compliance requirements. Usually the customer is good at what they do and they need several attendant services such as payroll and credit card processing. Both the outsourcer and the cloud service provider help the customer by taking over the management of this important service for them. When a service is outsourced, the organization loses their in-house expertise in these areas.

Both outsourcing and cloud computing introduce several risk management issues for the organization. In the outsourced model the outsourcer entirely performs the outsourced function. On the other hand, in the cloud computing model, the business simply uses the resources in the public or private cloud. The applications used or still the same but how they are used differ significantly. One of the key challenges in the cloud service model is the integration of various applications on the cloud. The business organization is ultimately responsible for all compliance requirements irrespective of the service being outsourced or moved to the cloud. Often the reason for outsourcing is the lack of expertise within the business to perform the business function. For example, a small banking institution needs to offer electronic banking for its customers. Otherwise, they would lose customers. In order to offer electronic banking the small bank needs to put in place highly secure practices and protect the servers through which the service is offered. This is cost prohibitive as well as the organization may not have the people resources to manage this high level of security. On the other hand the small bank has the ability to outsource the online banking aspect to a third party that specializes in that service. This is a viable business option. From the third party perspective, they are able to distribute the high cost of security among the many businesses that use their service. Since security is their primary focus they are able to keep up with all the advancements in technology in that field. The cost of this type of service is affordable for the small bank. However, if the small bank were to consider moving their online banking to the cloud, it still would be responsible for service management. The cloud provider only offers the resources. Hence, outsourcing is the preferred solution in this scenario.

Traditional outsourcing is highly customized and the IT services delivery is locked in for a specified period of time, usually 3 years. Because of customization and the service lock-in period there is very little innovation coming out of traditional outsourcing models. They meet the current needs of the business customer but do not explore alternatives to deliver the service cheaper or better. Cloud computing by design is standardized, which is at the other end of the spectrum for flexible design. The standardization enables the cloud customer to subscribe to a service in a few hours to days as opposed to months it takes to make an outsourcing arrangement to work. Since there is much competition in the cloud service, innovation is key to success from the service provider perspective. Traditional outsourcing is more people and asset based that has a 1-to-1 arrangement with the outsourcer. On the other hand, cloud computing is more standardized in its offerings, with the ability to use much of the services in an automated manner. By design, the cloud computing model is 1-to-many service through multi-tenancy and virtualization. Unlike traditional outsourcing, cloud computing follows the pay-as-you-go model. This means that the customer is not locked in for an extended period of time with the service provider. Another drawback of traditional outsourcing is that the solution does not scale quickly. The hallmark of cloud computing is its scalability. A Gartner Research study shows that businesses spend 53 % of their IT budget on outsourced services. This spending growth is at 4 % annually. Cloud computing has opened up several possibilities for businesses of all sizes to use high-end computing at an affordable cost and its growth rate is nearly three times that of traditional outsourcing. So, more businesses are using cloud computing because of the standardized offerings over the public cloud and the availability of commonly needed applications via the Software as a Service.

In certain sectors such as health care and finance the audit requirements are stringent. The business in one such sector that uses an outsourcer or a cloud service provider would still be responsible for providing the audit data for compliance purposes. In the case of the outsourcer the business will be able to obtain the necessary audit data for internal controls and data access controls. If a cloud service provider is used, the business will have to rely on the provider's compliance certifications and audit data. If such data does not address the specific requirements of the business, then the business will have great difficulty obtaining such data through a third party audit of the cloud service provider. From a practical perspective, as the provider of services to many customers, the cloud service provider cannot expose their systems to third party audit of all its customers. In such a case the cloud service model may not be the best solution for the business. The outsourcer model is more nimble compared to the cloud service model in this regard.

Businesses that enter into a relationship with either the outsourcer or the cloud service provider need assurance that at the expiration of the service contract all data pertaining to the business are deleted. Considering the cloud service model, data is backed up and replicated in multiple locations for data availability. The business has the option of choosing a data storage region on the cloud. However, the business is unaware of where the data gets backed up. For this reason, the business may not have processes in place that could verify the deletion of all data with the cloud service provider. On the other hand, the outsourcer will be able to provide such a verification because their functions are performed regionally. Moreover, in the cloud service model the service provisioning happens over the internet and so it is accessible for businesses globally (FFIEC 2014). The customer base is very large in the cloud service model compared to the outsourcer model. The execution of the outsourcer model still has a regional focus and the number of customers is much smaller compared to the cloud service model. For this reason, the outsourcer model has the advantage of providing the necessary verification that all data are deleted at the termination of contract.

Even though the outsourcer model and the cloud service model are two different types of services, there is a high probability that the outsourcer might use some form of cloud service. In this case the business needs to verify that the cloud service used by the outsourcer is a private cloud. This is feasible for the outsourcer because of the volume of service to multiple customers. The outsourcer will be able to provide suitable assurances and validation data to the customer concerning data security and privacy of information. From a compliance perspective the customer will be able to verify that there is data isolation in the private cloud in spite of server virtualization. Since all services in the private cloud are controlled by the single entity, the outsourcer, there is security and privacy guarantee in this model.

Cloud computing is considered to be disrupting the outsourcing industry. Data for 2013 reveals the IT outsourcing has increased. The disruptive nature of cloud computing is noticeable both in IT Outsourcing (ITO) and Business Process Outsourcing (BPO). In the case of ITO, the preference is for Infrastructure as a Service (IaaS) which gives greater control for the business in using the cloud services. For BPO, the preference is for Software as a Service (SaaS). The reason for this is that the SaaS market is highly mature in Email, CRM and HR (ISG 2014). Moreover, the ISG study shows that the ITO growth is significant over the past 3 years. Small and medium sized businesses are using ITO more. The leading SaaS providers in these markets are Microsoft and Google for Email, Salesforce for CRM and Workday for HR. According to the Deloitte Global Outsourcing and Insourcing Survey, IT Outsourcing is the most outsourced business function at 76%. The primary reason for ITO is the need to reduce costs. In this regard cloud computing is also a beneficial option (Deloitte 2012).

6.6 Summary

Cloud computing has a significant market penetration. Small and medium sized businesses find greater acceptance of the risks associated with cloud computing. Cloud computing's benefits to businesses are significant. Many of the benefits offer the businesses access to high-end computing services on demand. The availability of services and cost benefits enable the customer to choose the services needed. It is important to realize that there are many risks to the customer associated with the choice of cloud computing, along with the many benefits. To mitigate these risks the customer has to spend additional financial resources, which reduces the cost savings by switching to cloud from in-house computing. For this reason many large businesses are still hesitant to switch their services to the cloud. Some of the major benefits of demand elasticity and service availability of cloud computing allow large businesses to choose private cloud service. This also enhances security and privacy as well as affords the large business to control their computing environment. We discussed at length how cloud computing is considered as a variation on traditional outsourcing, which the businesses are accustomed to using. The trend in IT outsourcing points to the development of industrialized IT services that are offered through cloud computing. This trend is expected to raise the adoption of more cloud services.

6.7   Review Questions

1. What are the benefits of cloud computing to small and medium sized businesses?
2. What are the concerns of large businesses in using cloud computing?
3. Describe five benefits of cloud computing to businesses.
4. Risks faced by a business in cloud computing vary depending on their size. Explain how these risks differ. How could a cloud customer mitigate these risks?
5. How could a business negotiate a stronger SLA to mitigate risks?
6. In what ways the traditional outsourcing model and the cloud computing model are similar?
7. In what ways the traditional outsourcing model and the cloud computing model differ?
8. What is meant by Industrialized IT service and how this is expected to change the future of cloud computing?

References

Bezos, J. (2008). Animoto experience. http://www.youtube.com/watch?v=uIc-VB-ke9o. Accessed 30 Jan 2014.
Catteddu, D. (2010). Cloud computing: Benefits, risks and recommendations for information security. In C. Serrao, et al. (Eds.), *Web applications security*. Berlin: Springer.
CSA. (2014). Cloud security alliance guidelines. http://www.cloudsecurityalliance.org. Accessed 30 Jan 2014.
Cloud Industry Forum. (2014). Overview. http://cloudindustryforum.org. Accessed 5 Feb 2014.
Deloitte. (2012). Global outsourcing and insourcing survey. http://www.deloitte.com. Accessed 30 Jan 2014.
FFIEC. (2014). Federal financial institutions examination council. http://www.ffiec.gov/. Accessed 30 Jan 2014.
Gartner. (2013). Cloud computing will become the bulk of new IT spending, Gartner Symposium, October 21–24, Goa, India.
Gellman, R. (2009). Privacy in the clouds: Risks to privacy and confidentiality from cloud computing. *World Privacy Forum* http://www.worldprivacyforum.org/2011/11/resource-page-cloud-privacy/. Accessed 1 Oct 2014.
Interxion. (2013). The evolution of the European cloud market, Whitepaper. http://www.interxion.com/. Accessed 1 Oct 2014.
ISG. (2014). Information services group. http://www.isg-one.com/. Accessed 30 Jan 2014.
Kopytoff, V. (2012). The cloud carries risks too. Bloomberg Business Week, August 7.
McAfee. (2014). Pledge software token. http://www.mcafee.com. Accessed 30 Jan 2014.
NIST. (2013). Guide to Attribute Based Access Control (ABAC) definition and considerations, SP 800-162. http://csrc.nist.gov/projects/abac/july2013_workshop/july2013_abac_workshop_abac-sp.pdf. Accessed 1 Oct 2014.
NSBA. (2013). National small business association survey.

Open Stack. (2011). Open stack overview. http://www.openstack.org. Accessed 30 Jan 2014.

Ponemon Institute. (2011). Security of cloud computing providers study. Ponemon Institute, May.

Research in Action. (2013). Hidden costs of managing applications in the cloud. Whitepaper. http://www.compuware.com/content/dam/compuware/apm/assets/whitepapers/WP_Costof-Cloud.pdf. Accessed 1 Oct 2014.

Srinivasan, S. (2014). Risk management in the cloud and cloud outages, Chapter 8. In S. Srinivasan (Eds.), *Security, trust, and regulatory aspects of cloud computing in business environments*. Hershey: IGI Global.

US-CERT. (2012). Common risks of using business apps in the cloud. http://www.us-cert.gov/sites/default/files/publications/using-cloud-apps-for-business.pdf. Accessed 30 Jan 2014.

Weinman, J. (2012). *Cloudonomics: The business value of cloud computing*. New York: Wiley.

Chapter 7
Hidden Aspects of a Cloud Computing Contract

Abstract Cloud service providers offer a contract to their customers. The contracts do not deviate much from one another. The cloud service provider prefers the contract to be executed as a click-through online. In a way the cloud service providers keep the contract standard without customization. There are two main reasons for the standard contract. First, the services are standardized and the customer has the full range of services to look at and decide on the services that they want. Second, with numerous customers it will be impossible to manage customized contracts that differ from one another. Review of the standard contract reveals that the Service Level Agreement (SLA) does not give the customer much recourse. Invariably the SLAs favor the service provider. There is a new trend among cloud service providers and customers to seek insurance to mitigate the risks. In this chapter we look at the many hidden aspects of a cloud contract and point out where clarity is needed. Cloud service providers have privileged users who have the potential to access customer data on the cloud. The standard contract does not specify how privileged users will be monitored. Also important to the customer is the ability to switch service providers and move data across providers easily. This requires the service provider to have the resources to support high speed data transfer in a standardized format. We will look at the provisions made in the contract for such a move.

Keywords Cloud computing · Contract · Log data · Uptime · Privileged users · Data portability · Insurance

7.1 Introduction

Cloud computing is evolving rapidly as an effective alternative to in-house computing. The cost advantages and service availability make it an attractive alternative for many small and medium-sized businesses. Since cloud services are provided by a third party every business needs a contract to codify all the service guarantees of the service provider. Many businesses consider a variation on outsourcing. Typically in outsourcing the contract is customized to the customer needs. Moreover, it takes a significant time before the outsourced service becomes available to the customer. In the cloud service model the customer is able to select from a set of standardized services available over the internet and select a cloud service provider. The service becomes operational within a few hours or at most a few days. Because of the stan-

S. Srinivasan, *Cloud Computing Basics,* SpringerBriefs in Electrical and Computer Engineering, DOI 10.1007/978-1-4614-7699-3_7,
© Springer Science+Business Media New York 2014

dardized nature of the service the cloud service providers offer a standard contract. A review of several contracts shows that there is not much variation among the contracts. More striking is the fact that many of the provisions in a cloud contract favor the service provider, with very little remedy for the cloud customer. A cloud customer chooses the cloud service based on the fact that the service provider has the advantage of economies of scale and so they are in a better position to offer an affordable service that is up virtually all the time. Small and medium-sized businesses will have great difficulty managing their own computing system that is up virtually up all the time, let alone the cost factors. Large businesses, on the other hand, have many resources available and so they could use their own computing system and supplement that with additional resources from a cloud service provider.

When a business chooses cloud sourcing they want the service to be reliable and available nearly all the time. To support this feature the cloud service providers offer a service uptime guarantee of 99.95 %. This means that the service could be down only for 21 min a month in order to have such a high level of up time. This level of service guarantee is typical from all the major cloud service providers such as Amazon Web Services (AWS), Microsoft, Google, Rackspace and Salesforce. All these service providers had service outages over the last 5 years. On many of these outages the service was unavailable for several days. This clearly was not in line with the service guarantee. Typically service contracts have provisions that when a service guarantee is not honored then the service provider incurs a penalty to compensate the customer for lack of service. In the case of cloud service contracts, because there is so little variation from the standard contract, the cloud service provider includes provisions that will protect them from paying any penalty to the customer. The cloud contract specifies that in order for the service guarantee to be honored the cloud customer must choose multiple Availability Zones (AZs). An Availability Zone contains both computing resources such as servers, applications and storage devices. Each service provider divides their service into multiple regions around the world and places several AZs in each region. This is more of a requirement from a legal perspective because many foreign governments prohibit storage of data of their citizens outside their borders. The common regions are North America (US, Canada, Mexico), Latin America, Europe-Middle East-Africa (EMEA), and Asia-Pacific (APAC).

We mentioned above certain standard regions. This grouping of countries by region is by no means accepted by all cloud service providers. There are several variations in the countries included in a region based on the service focus of the cloud service provider. Because of the constraints of USA PATRIOT Act and the recently disclosed NSA data surveillance, many foreign governments do not want their country data in regions accessible to the US Government. Europe has specific Safe Harbor provisions that protect the data privacy of their citizens (Safe Harbor-EU 2000). Within Europe, Switzerland has some additional Safe Harbor requirements. These are codified in a separate Safe Harbor Agreement with Switzerland (Safe Harbor-Swiss 2008). In situations described above the cloud customer may need modification of contract to reflect that data privacy concerns will be accommodated. According to Hon, Millard and Walden, most cloud service providers are not willing to modify their contracts for lower end customers on the level of service used (Hon 2012).

The primary purpose of seeking changes in the contract is to reduce the risk for the business. Another trend emerging as an alternative to contract modification is insurance. From the cloud customer perspective the insurance seeks to mitigate the damage caused by loss of data, security breach or service availability. Hon et al. point out that in many cases the cloud providers are seeking insurance against customer claims for service disruption. Thus, a new service industry is emerging to provide insurance service to the cloud service providers and customers. Use of such a service adds to the cost of using cloud service to the customer which in many cases would nullify the cost savings. Insurers insist on certain compliance certifications before offering their service. Also, the cloud service providers are willing to cooperate with the insurance carriers in the form of allowing them to gather the data they need on the service management aspect. This is noteworthy because the cloud service providers are not willing to allow third party audit of their services at customer request.

It is a fact that most cloud service providers are unwilling to modify their standard contract. The customer ends up with a standard contract that is generated through a process called "click-through" during service setup. Getting a contract through "click-through" is considered a sign that the industry has not matured enough to negotiate specific contracts. This general thinking will not be applicable to cloud services because of their design to offer utility computing with easy access to their services. Moreover, in order to provide cost advantage to the customer many cloud service providers lack the legal resources to have modified contracts. A survey in UK by the Cloud Industry Forum showed that nearly half of all cloud contracts were not customized (Cloud Industry Forum 2011).

In addition to cloud service providers there are two other groups that are involved in contracts. The first group is Cloud Service Brokers (CSBs). The CSB is a significant force in the use of cloud service, especially for small and medium-sized businesses (SMBs). According to Forrester Research and MarketsandMarkets study, the CSB service accounts for over $ 10 billion of IT spending for the cloud customers. They have an important role to play in that they look out for features that are essential in cloud service and advise cloud customers accordingly. Among the SMBs the demand is more for SaaS applications whereas for the larger businesses the need is in Infrastructure as a Service. Moreover, as a third party, the CSBs are able to find better service rates because of the volume that they can generate among the many cloud customers that they have. The CSBs are investing in CapEx to provide services to customers who they expect to use their CapEx investment as OpEx for them. The second group is Cloud Service Integrators (CSIs). A CSI offers service integration features for the cloud customer which goes a step further than the advice the CSB typically offers. CSI services are value added for the cloud customer because they can integrate the back office work and identity management with cloud service. In a way the Cloud Service Brokers are emerging as Cloud Service Integrators. Integrators are themselves cloud service providers to SMBs, especially SaaS applications. The integrators take on some risks in order to retain some customers. Major telcos such as Verizon and AT&T who have a large customer base are trying to benefit from the CSBs who can market SaaS solutions over the telco infrastructure. Two Cloud Service Integrators are HyperStratus and Appirio.

7.2   Service Level Agreement

Any organization that signs a contract with a service provider should codify the service guarantees through a service contract. The Service Level Agreement (SLA) is the common instrument for this. Service providers make SLAs favorable to them and add constraints that are not possible in software licensing. Such constraints are added for increased revenue. For example, Microsoft Office 365 cloud service restricts the number of emails that a cloud customer could receive in a day. Exceeding this daily limit would incur cost for the customer. Given the demand for cloud services many cloud service providers are not willing to change their standard terms for SLAs even for large companies and Cloud Service Integrators. An exception occurs where large government entities and financial institutions with compliance requirements negotiate on SLA terms. In such cases the volume of business combined with a niche service area attracts the cloud customer to agree to changes for the standard terms in their SLAs. Clearly SLAs apply only for paid services. It is common practice in many businesses where customers sign up for free cloud services with click-through standard terms. Invariably these free services lead to other paid services where SLAs are needed. Since the relationship with the provider already existed it will be difficult to negotiate any favorable terms in the SLAs for the customer.

In SLAs the common expectation is for the provider and customer to limit their liabilities. The providers are unwilling to increase their liabilities through SLAs. Global customers have a better chance of limiting their liabilities through SLAs but the service provider usually accepts only "direct losses" to the customer because "indirect losses" may be difficult to verify (Cloud UK 2011). Further, the service provider places a cap on the liability limited to a certain percentage of the fees paid for the service for a specific period like 1 year. The reason for this approach is that most of the services are commoditized. Also, focus on SLAs may not provide the necessary protection when the provider may not be credit worthy. It is standard practice with cloud service providers to have redundant backups of data. However, the provider may not be willing to guarantee through an SLA the data availability or integrity. It is usually offered as an add-on service for a fee. Because of this fact many customers may not insist on modification to the SLA for data availability. One common concern with cloud service is security. To overcome this concern the service provider assumes responsibility for data confidentiality violations. However, data breach is not considered a data confidentiality violation. Since data breaches are more likely to occur than unauthorized access, this aspect of SLA favors the service provider.

Cloud service providers emphasize that protecting data and service on the cloud requires shared responsibility with customers. The cloud customer using IaaS or

PaaS service has greater control over their cloud service than a SaaS customer. For this reason, the cloud service SLA for IaaS or PaaS need not focus on data backup, recovery, security or access control. For SaaS customers this will be difficult to incorporate in an SLA because of the commoditized nature of the service at a low cost. The common mitigation strategy for SaaS service users is the backup and recovery service of the integrator who also guarantees data integrity given the relatively smaller size of the customer base for such a service when compared to the cloud service provider (Geyer and McLellan 2011). Service availability is a key metric for businesses of all types but large businesses pay more attention to this metric than small and medium-sized businesses. One of the major benefits of a cloud service is its very high availability. Yet, when the service is disrupted even for a short while, its impact is huge because of the very large customer base that is global in nature. Cloud service providers are unwilling to modify their standard SLA for availability because they offer the service as a commodity. A major SaaS service provider like Salesforce does not offer an SLA and 99.8 % of its customers are not particular about an SLA. Regarding service availability, its customers expect only 99.7 % availability and Salesforce offering has a higher availability rate than 99.7 %. For these reasons, it will be difficult to have a customized SLA for any cloud service regarding service availability. Moreover, companies like Amazon Web Services (AWS) which offer an SLA with service availability insist that SLA provisions will be valid only when a customer contracts for service in more than one Availability Zone. This indirectly implies that the customer backup their data in more than one Availability Zone. AWS service availability data indicates that they do not have concurrent failures in multiple availability zones. AWS has also defined that their service availability should be measured as a percentage over a period of 1 year rather than on a month to month basis. Thus, a 99.95 % availability of their service which only allows for 21 min of downtime in a month becomes over 4 h of downtime spread over 1 year. These changes built-into the provider SLAs make them favorable to the service provider.

In mission-critical services the customer wants low latency meaning faster response time. Virtually all service providers such as Amazon, Google, Microsoft, Rackspace and Salesforce all depend on a telco to provide connectivity. Since they do not control the actual communication network they are unwilling to guarantee low latency through an SLA. However, a cloud service provider such as Terremark, which is owned by the telco Verizon, offers low latency guarantee. Terremark owns the NAP of the Americas data center in Miami that provides protection against natural and man-made disasters. A classic example of high availability, reliability and security through Terremark cloud service is the Subway Cash Card management program since 2006. The mission-critical nature of this program in North America for Subway had maintained 100 % availability through Verizon and Terremark. In the case of other cloud service providers who are unable to provide an SLA, some service integrators have assumed the risk for a fee and are providing low latency guarantee via their own SLA to the customer. Hence, the cloud customer may have the option to get an SLA with some of their concerns addressed through an integrator rather than the service provider.

One of the hidden facts about the efficacy of an SLA is its use in the event of a perceived contractual violation. The cloud service providers want to link the SLA clauses to parts of their web site that they control and which they amend periodically without informing the customers. Thus, if the customer had a feature negotiated with the service provider then it could be nullified because of its linkage to other clauses in the company web site that they could modify to suit their needs. From an enforcement perspective, in the event of an SLA remedy, the customer only gets service credit based on what they paid for a specific service that was violated. Since the payment for such a service is small on a monthly basis, even if a credit were to be forthcoming, it would not offset for the customer's reputation loss, because of which they may lose customers of their own if they happen to provide a service. Because of this aspect of an SLA some large cloud customers simply do not count on the SLA to provide any remedy, instead they pay for insurance to cover their risks. An SLA requirement for the customer could be transparency in the system performance, incident handling and management practices. After several well-publicized outages over the past 3 years many cloud service providers are now providing system performance data on their web site that anyone could monitor. For example, the AWS system performance web site is called AWS Service Health Dashboard (AWS 2014), Google's corresponding web site is Apps Status Dashboard (Google 2014), Window's Azure site is Windows Azure Service Dashboard (Microsoft 2014), Rackspace site is System Status (Rackspace 2014), and Salesforce site is Analytics (Salesforce 2014).

Many cloud customers consider it important for them to know where their data is stored and processed. This is a compliance requirement for many customers. For this reason the cloud service provider makes available the location of their data centers for customers to choose from when contracting for cloud service. Even though this location information is guaranteed by the service provider in the contract, the provider has statements reserving the right to use third party sub-providers. Such sub-providers may not be in the area where the data center is located. Consequently, some customer data might be transferred to the sub-provider outside of the area designated in the contract. Cloud service providers do not consider this a violation of the contract and often the customer is unaware of the exact location of their data. This ambiguity is typical of the standard click-through contract. Of all the cloud service providers only Salesforce has taken the extra step to give the ability for the customers to verify where their data is stored and processed by using the web site trust.salesforce.com. Many global cloud service providers in US point out that the US-Europe and US-Swiss Safe Harbor Agreements provide the necessary legal protection for customer compliance. For these reasons, the cloud service provider does not modify their standard contract.

Security is considered the top concern by many potential cloud customers across multiple surveys (Ponemon Institute 2011; DLA Piper 2012). Cloud providers note that those customers who use their Infrastructure as a Service and Platform as a Service have full control over the cloud infrastructure they use and so they are responsible for any security measures needed. In the case of Software as a Service the cloud providers note that they have control over the infrastructure. When such customers need modification of the standard contract to reflect their security needs, the cloud providers resist change of contract. The primary reason cited by cloud providers is that too much transparency in their security practices is in itself a security threat. For this reason they do not support audit of their security infrastructure by cloud customers. This argument has some validity. However, the cloud providers could facilitate integration of more security solutions with their offerings such as one time passwords. Since this would increase the cost of service the cloud providers offer the security solutions for a fee. Another solution offered by cloud providers for security practices transparency is their compliance with ISO 27001 Standard. In lieu of SLA modification the cloud customer could be assured of security practices that comply with Cloud Security Alliance Security Standard, Cloud Industry Forum Standard and Open Data Center Alliance Standard (CSA 2011; Cloud Industry Forum 2011; ODCA 2011).

Businesses consider an SLA as important to expect certain level of service from the provider. It is standard practice for businesses to negotiate an SLA that meets their service expectations. With this goal in mind the legal unit of the business may want to get involved in negotiating an SLA. When it comes to cloud service this model does not work for several reasons. Cloud service is offered as a utility service. The service is available globally to thousands of customers. The customer base is often very large and diverse. One of the defining features of cloud service is quick deployment. Goal of cloud computing is to keep the cost of service low in order to attract more businesses to subscribe to the cloud service. In order to keep the cost down the cloud service providers do not have a large legal unit that can review unique expectations from its cloud customers. For all these reasons the cloud service provider offers only a standard contract that is available online for click-through acceptance. Even though a customized SLA is not the norm in cloud service, the cloud customer should expect certain details spelled out in the contract. One requirement should be to know how and when a contract would terminate. Are there provisions for an abnormal termination of contract? As and when the contract terminates how will the cloud customer get access to their data and support for transfer of data to another source? These are important aspects that must be addressed in a cloud contract, even a standard contract.

Our discussion shows that cloud customers value an SLA. However, cloud service providers do not want to offer any customized SLAs. A quick review of the SLAs from AWS, Windows Azure, Google Apps and Rackspace will help understand the main features that service providers support through an SLA. Amazon Web Services (AWS) is the largest cloud service provider. AWS offers SaaS, PaaS and IaaS service types as well as public and private cloud deployment models. AWS service uptime is at 99.9%. Popular AWS service components are Elastic Compute Cloud (EC2), Elastic Block Storage (EBS) and Simple Storage Service (S3). AWS data centers are distributed around the world. The data centers are grouped into Availability Zones (AZ) and several AZs form a region. AWS' SLA requires the customer to sign up for two AZs in order for the service uptime in the SLA to be honored. The service uptime is measured by spreading it over a 12 month period instead of on a month to month basis. This gives AWS the ability to average out their total downtime in any particular month. In the event of a claim, the period of measurement starts after the last claim. AWS SLA does not make it clear whether they exclude their routine service maintenance times for SLA purpose. Since they stagger their routine maintenance times by Availability Zones it would appear that they do not exclude the service maintenance times in their SLA, because the SLA remedy would apply only for customers who sign up for multiple Availability Zones. It is the responsibility of the customer to initiate any SLA remedy and that too follow a strict notification time period. AWS offers a 10% service credit when the service uptime falls below 99.9% and a 25% service credit if the service uptime falls below 99%. All service credits are applicable for a future service period. AWS offers a CloudFront and CloudTrail service that would help the customer (AWS CloudFront 2014; AWS CloudTrail 2013).

Windows Azure offers a 99.95% service uptime guarantee. Their service uptime guarantee includes the time that their system will be down for maintenance. Since Windows Azure requires the use of multiple data centers for SLA remedy to apply it is reasonable to expect that their routine maintenance downtime will be included in the calculation of their uptime for SLA purpose. The uptime calculation is based on the billing month for each customer. It is the responsibility of the customer to notify Azure of their claim because of service unavailability. Azure calculates the service credit at 10% for service availability below 99.95% and at 25% credit for service availability below 99%. The service credit could be applied to the current billing period itself. Google Apps provides a service uptime guarantee of 99.9%. It is not clear from their SLA whether they require the use of multiple data centers by the customers before they can claim remedy from the SLA for uptime. They exclude the routine maintenance downtime in calculating their service uptime. This is an indication that they do not require commitment to multiple data centers by the customers. Google Apps calculates the service uptime on a calendar month basis. The customer is responsible for notifying Google Apps of any service disruption to claim remedy. The service credit is applied by the number of days of service on a sliding scale. If the service uptime is less than 99.9% then a 3-day service credit applies. For service uptime less than 99%, a 7-day service credit applies. If the service uptime falls below 95% of the time, a 15-day service credit applies. The credit is applied to the current billing period. Rackspace is the only company that claims that they guarantee service at 100%. A closer look at their SLA shows that the exclude their routine service maintenance downtimes from the uptime calculation. The monthly service downtime is around 1 h and so it changes the effective uptime from 100 to 99.86%. The service guarantee is calculated on a billing month basis. It is the customer responsibility to notify of any service outage in order to claim SLA remedy. Rackspace uses a flat 5% service credit for every 30 min of downtime. The customer is eligible up to 100% of service credit if the service becomes unavailable for a total of 10 h during a billing month. The credit is applied to the current billing month. We summarize in Table 7.1 the various features described above in connection with the Service Level Agreement and its enforcement.

We conclude this section with one important obligation of the cloud service provider to the customer that is part of the SLA. Often the cloud customers may have users of their own for services that they offer. In this case the cloud customer does

Table 7.1 Summary of SLA features from major cloud service providers

SLA feature	AWS	Windows Azure	Google apps	Rackspace
Service uptime guarantee (%)	99.9	99.95	99.9	100
Routine server maintenance downtime included or excluded from uptime calculation	Unclear	Included	Excluded	Excluded
Use of multiple data centers required for guarantee	Yes	Yes	Unclear	Unclear
How is service guarantee calculated?	On a 365 day basis or since the last claim	Billing month	Calendar month	Billing month
Responsible party for reporting violation	Customer	Customer	Customer	Customer
Level of service credit	10% for SA<99.95% 25% for SA<99%	10% for SA<99.95% 25% for SA<99%	3 day credit for SA<99.9% 7 day credit for SA<99% 15 day credit for SA<95%	5% for 30 mts downtime Maxm. 100%
Credit applied for future service only	Yes	No	No	No

Note: *SA* Service Availability

not control the infrastructure but is responsible for e-discovery. This means that the service provider should be able to create a snapshot of all aspects of the service, especially communications, and store it for e-discovery. The cloud service contract may not specify details on e-discovery but the customer should be able to create the snapshot with the help of the service provider so that the authenticity of the snapshot is vouched by the service provider.

7.3   Sharing Log Data

Cloud service providers gather plenty of data such as user login details to a virtual server, server status update, server maintenance update and operating system patch updates, etc. Much of this information helps the service provider to meet their compliance certification requirements for audit standards such as SAS 70. The customer will also require some of these log data to meet their compliance requirements. This is one aspect of log data that helps with compliance. Other uses for log data pertain to security and real time processing of related data. From the security perspective the cloud customer would want access logs of all attempts to reach their virtual servers. This should include cloud service management access as well. From a customer perspective having access to this data on demand enables them to be assured of their security. Technology exists today that will help gather this type of access data. Cloud service providers should make available such data proactively and leave it up to the customer to process such data as they need it. This has the added advantage of providing system uptime statistics to the cloud customer. A third type of log data gathering by the cloud provider is based on certain triggers. This will require the ability of the log tool to work across multiple applications. For example, as part of video surveillance a business gathers continuous data feed from a camera mounted to monitor people movement. In this context the log data is gathered from the video stream and processed to identify individuals. Since the data stream for the most part consists of routine data the need arises to pick out only special events for further analysis. The organization may have a watch list for certain people based on the nature of work they do. When one such individual is recognized from the video stream then it would trigger a series of snapshots of data that are assembled in one source. In order to process such log data the preferred data format is JSON (JavaScript Object Notation). JSON is a light-weight data interchange format that can be used by machines to parse and generate data that can be read by humans (JSON 2014).

As part of innovative applications that are possible with cloud service, cloud providers are beginning to use log data to derive useful information. Amazon Web Services (AWS) has introduced CloudTrail to log all API calls to AWS services. This log data arranged in JSON format is stored periodically in the customer's S3 storage bucket. AWS is offering CloudTrail as a free service to enhance customer trust and to offer a tool to the cloud customers to meet their compliance requirements. Global standards such as ISO 27001 and PCI DSS have data logging requirements to validate security practices of the implementing organizations. CloudTrail evolved from gathering log data for several compliance standards. Other cloud service providers also offer similar logs for customer manipulation. In the case of Microsoft the log data compiler is the Performance Monitor. The Performance Monitor can alert the customer when certain events occur. Google provides log monitoring using Logs API. The Logs API gathers all requests to apps and stores them in a log file. It is important to realize that log data will grow very rapidly. Google provides 100 MB free data storage for log data and charges for storage beyond that limit. This is similar to AWS practice where customers pay for S3 storage use and AWS stores log data in S3 buckets. Salesforce also facilitates gathering log data for customers through its debug log.

From the descriptions above it is clear that log data is available for cloud customers to process. By its very nature this type of log data for cloud services will fall into the category of Big Data. In order to benefit from such log data the cloud customer must analyze and extract relevant information quickly. Even though the log data is designed to be stored for 1 year, the main use occurs close to the time of data capture. To facilitate processing large volumes of data several application providers have emerged. Some of these application providers are Logentries, Sumo Logic, Loggly, Splunk and PaperTrail. These third party application providers have pre-built APIs that work with tools like CloudTrail and analyze user behavior to detect unusual usage patterns. These useful tools first gather large volumes of data from multiple system sources and reduce them to a small set for further analysis using Big Data Analytics toolset. These tools use pattern recognition to classify the data into three risk categories—high, medium and low. The system administrators then focus on the log data based on the risk classification.

The importance of sharing log data by the cloud service providers with the cloud customers is further reinforced by the positive benefits to the customers. Financial services companies such as large banks generate terabytes of data from several security data points daily. When such data are stored for an extended period of time such as 1 year, the volume of data grows to petabyte scale. The purpose of such extended storage is to infer long-term correlations. To process such large data volumes with low latency searches the National Security Agency (NSA) released the tool Accumulo through Apache open source. These developments show the government interest in making tools available as open source for benefiting from processing large volumes of log data.

The data logs originate with the cloud service provider. In a recent study Sundareswaran, Squicciarini and Lin show how data on the cloud can be protected by timely access to log data. Their approach takes advantage of a feature of JAR (Java ARchives) which is connected to the file in question before it is uploaded to the cloud. The JAR file has the access controls specified for accessing the target file and it automatically logs all information about the user seeking access to the target file. Periodically the log data is sent to the target file owner who can verify that the access control mechanisms specified in the JAR file are honored. In this scenario the cloud service provider facilitates the functioning of the JAR file in the virtual machine of the cloud. The requirements are minimal, the VM needs a Java virtual machine installed. Thus, the user is able to benefit by the shared log data from the cloud (Sundareswaran 2012).

There is another form of capturing user input in the cloud, which is similar to log data. In this context the log data refers to data that is captured automatically based on some action by a user such as accessing an application on the cloud. The proposed variation in this respect also captures user data. This project from the Massachusetts Institute of Technology (MIT) is called Cloud Data. The Cloud Data project gives a mechanism on the cloud to capture user interaction. It has the potential to capture trigger data based on specific actions by a user (MIT 2014). Another similar cloud resource is by Nimbits. This is a free application that runs on the cloud as a Platform as a Service (PaaS). Developers can use Nimbits to develop applications that are triggered by certain events such as a sensor data trigger or stock market action trigger. In this application the cloud data based on log data or location data triggers new events which could result in a cascading series of actions (Nimbits 2014).

7.4 Service Uptime Guarantee

Cloud services are known for their high reliability and service availability. If a cloud customer were to host the service in-house they will face significant challenges in maintaining the same level of availability. Yet, cloud customers expect a cloud service to be available 100% of the time. Such an expectation is unreasonable because technology is bound to fail sooner or later. In this section we look at service uptime guarantees by major cloud service providers and how well they are honored. For this analysis we looked at Amazon Web Services (AWS), Microsoft, Google, Rackspace, VMware, and Terremark. All these service providers guarantee a service uptime of 99.9%. This translates to a maximum downtime of 53 min per month. Our goal in this section is to highlight two things—how service providers try hard to provide a high level of service uptime and what factors hinder such effort.

Of the six major cloud service providers, only Terremark is backed by a telecommunications provider. Terremark is a division of Verizon Communications, which has a large communications infrastructure that they control. All other cloud service providers are dependent on telcos for the network part of the cloud service. Cloud companies invest heavily to provide reliable cloud service. The infrastructure investments by some of these companies are as follows:

AWS	$ 12 billion
Google	$ 21 billion
Microsoft	$ 18 billion

All companies have created new data centers closer to the locations of their customers to provide low-latency service. In spite of their best efforts outages are typical in cloud service. Over the past 5 years there have been several well publicized cloud outages from AWS, Google and Microsoft. Other cloud service providers are not immune to cloud outages. Still these companies have been able to provide a very high uptime for their services.

Cloud service providers realize that service uptime builds trust of their customers. Since they provide service to thousands of cloud customers they do not want to be held liable should their service uptime fall below their guarantee. Their primary reasoning for this attitude is that they are providing a commoditized service at a low cost. Many cloud service providers have started interpreting their service uptime guarantee to be measured against a whole year instead of month by month. Thus, a 99.9% service uptime guarantee would translate to 53 min × 12 months = 636 min per year. The service providers feel that given the over 10 h window per year for downtime is something that they could strive for in their service. Another standard exclusion in the uptime guarantees is the routine maintenance time, which is typically 1 h every month. Given these caveats, a 100% uptime guarantee really translates to 99.86% availability (Baset 2012). Based on the major outages that lasted anywhere from 30 h to 3 days the cloud service companies would still be liable for lack of service over an extended period of time. Experience shows that in many cases the service outages have been localized and so the service providers have tied their service uptime guarantee to the use of their distributed data centers. Another typical exclusion for service uptime calculation is that the service downtime at any time lasts more than 10 min. For example, AWS service uptime guarantee applies only if the customer subscribes to at least two of their Availability Zones. By tying their service guarantee to multiple Availability Zones their focus is on making sure that multiple Availability Zones do not go down simultaneously.

Businesses that seek a service guarantee from cloud service providers are trying to protect their reputation. So the natural remedy for not meeting the service guarantee is some form of penalty. All cloud service providers only offer service credit for any violation of service guarantee. The service credits are anywhere from 5 to 100 % of the fee paid for the service or in some cases a multiple of the service fee (CloudSigma 2014). In any case the financial value of such service credits to not offset the reputation loss for businesses and so many large businesses do not concern themselves about service guarantees, instead they opt for insurance coverage for any service loss. At this time the cloud insurance is in its nascent stages.

In the above paragraphs we have highlighted what service uptime guarantee really means to the customer. Cloud customers automatically tie service availability to data availability. From the customer perspective the important thing is data availability. This is one reason why cloud service providers are providing uptime guarantee only when multiple data centers are used for storage. Often data gets corrupted due to software bugs and accidental data deletions. So, real time disk mirroring would only store the erroneous data in its multiple copies. The real solution is to do point-in-time backups or snapshots. Cloud storage would still be reliable because of redundancies built-into cloud storage. It is up to the customer to create the snapshots and store them periodically.

Businesses want low latency in delivery of any service using the cloud. Associated with this need is the data availability for delivery to the customer. Cloud service providers handle this need by using several edge servers around the globe. These edge servers constitute the Content Delivery Network (CDN). CDNs can handle static, dynamic and mobile content. Also, CDNs can deliver music, video and games. Cloud service providers develop CDNs so that the content is located in a server closest to where the customer is located. The CDN architecture replicates data in different edge servers based on need. In this sense the CDN supports high data availability even when some cloud services are down in some regions. AWS' CDN offering is known as CloudFront (AWS CkoudFront 2014).

7.5 Privileged Users

Cloud service providers control the computing infrastructure. As part of the management functions certain employees of the service provider will have to have elevated privileges. We identify such users as privileged users. The role of the privileged users is to monitor that all virtual machines are functioning properly and the applications running on the various virtual machines stay within their allotted memory areas. The cloud customers are responsible for the data that they process and store in the virtual machines. The privileged users monitor that the applications used by customers stay within their boundaries. However, since they have access to customer data there should be appropriate logging of all access to customer data. In order to build customer trust the service provider should limit the number of privileged users to a handful. Moreover, the service provider should identify their policies and practices with respect to the hiring of privileged users.

Privileged users access customer data files usually for restoration, if necessary. Otherwise the customer files are not looked at by privileged users. This is the norm. However, a rogue privileged user could have an unauthorized access to customer data. In such extraordinary situations the customer should be able to note such access from an automated access control log that is read only and cannot be deleted by privileged users. Another way to protect content of customer data is through encryption, tokenization, audit and monitoring. Each of these aspects involves additional cost to the customer. Since customers are ultimately responsible for integrity of their data they may have third party monitors continuously monitor all database access to their data. This includes privileged user access. If a rogue privileged user deactivates the notification provision of the monitoring activity then that should trigger an alert to the monitoring entity. Encryption appears to be an easier solution because the customer can hold the encryption key and store the encrypted data in the cloud. However, the encrypted key will be needed to search the database and managing this requirement may not be easy. Tokenization is another approach in which authorized users will be able to obtain a secure token from the customer and then access the database. The idea of these approaches is to have access control to the data. This will be needed in case the customer has compliance requirements with respect to Sarbanes-Oxley Act, Health Insurance Portability and Accountability Act or Payment Card Industry Data Security Standard compliance.

Privileged users have unfettered access to the cloud's physical infrastructure as well as the software on those servers for maintenance purposes. Because of the diverse nature of the work to be performed there are multiple groups of privileged users at the cloud provider. To protect against insider attacks on the cloud content Bleikertz, Kurmus, Nagy and Schunter discuss five different areas to focus on. We summarize this work to show how cloud customers should be aware of possible insider attacks on their data. The potential for malicious behavior arises during maintenance when many manual processes are involved, including replacement of the hardware. For this reason the cloud management practice should include separation of duties among the various groups of privileged users. The three main groups involved are the hardware team, the security team and the remote maintenance personnel. Organizational policies and access control to physical infrastructure provide audit trail for the first two groups of users. For the group of remote maintainers the principle of least privileges should be the norm (Bleikertz 2012). As the need arises for higher privileges these administrators can elevate their privileges, which are automatically recorded in a store-only device. This storage content should be accessible to customer initiated audits.

Focusing further on the privileged users and the possibility of insider attacks during routine maintenance, any modifications to the system content should be visible

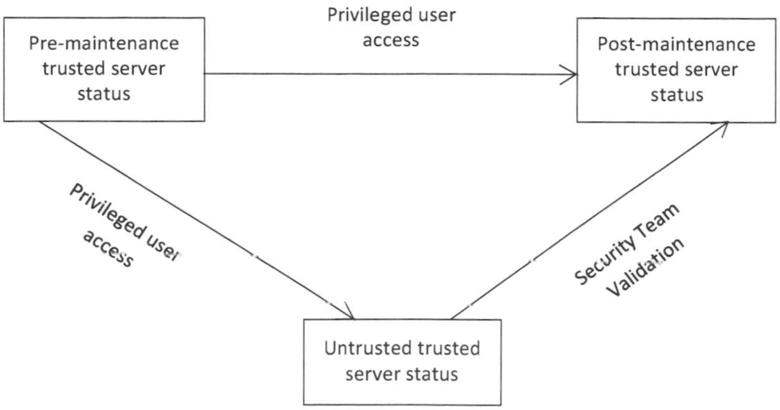

Fig. 7.1 Customer trust and privileged user access

to the security team. To prevent insider modification of customer data resulting in loss of integrity, the security team should be able to roll back the modifications. The security policy of the cloud service provider should move a physical hardware from trusted to untrusted mode until it is verified by the security team and moved into the trusted mode. This security procedure should be communicated to the cloud customers to enhance customer trust. Figure 7.1 illustrates this process. The nature of maintenance requires the privileged users to be able to view the access logs. However, in order to maintain data confidentiality the privileged user should not be able to access the customer data unless authorized by the customer. The reason for such a need is that certain problems connected with corrupted data is unverifiable to identify the cause in a test VM or the problem disappears when the VM is turned off. This is an indication that some malicious code has infected the customer application and is embedded. In order to detect and remove such malicious code the VM must be running and the privileged user should be able to access the VM during run time. Thus there are instances when privileged users may access customer data with the knowledge of the customer.

As an enhancement of this process it is possible to monitor privileged user access to customer VMs in light of the process described in Fig. 7.1. In this approach the privileged user access to customer VMs is layered. In Layer 1, the privileged user has no access to customer VM but can maintain the physical infrastructure in which the VM resides. In Layer 2, the privileged user has Read Only access to customer VM for checking the maintenance logs. In Layer 3, the privileged user is granted "White-Listed Write Access" to customer VM implying that all non-sensitive system files can be modified, such as applying routine patches. In Layer 4, the privileged user access is at the "Black-Listed Write Access" level meaning that any write at this level should place the VM in an unsecure state. As mentioned in Fig. 7.1 the VM has to be reviewed by the security team and placed in the Post-maintenance trusted state. Black-Listed write access would include changes to the operating system and addition of new software. Finally, in Layer 5 the privileged user is granted full access to the customer VM and it should automatically place the customer VM in untrusted mode until verified and approved by the security team (Bleikertz 2012).

7.6   Data Portability

Cloud service growth is exponential. There are less than 50 large global cloud service providers whereas there are several hundred smaller cloud service providers who use Infrastructure as a Service feature from these large cloud service providers and offer cloud service of their own. By design the large service providers encourage such use of their cloud service. When so many cloud service providers are in the marketplace, it is inevitable that some will go out of business or the cloud customer may want to move their service to another cloud service provider. In either case the customer data needs to be moved from one provider to another within a limited period of time. Support for this feature in cloud service is called data portability. At this time there are no global standards to support data portability, which essentially supports the ability of cloud customer to change the service provider easily. Since the cloud computing industry is still maturing it will take some more years before the customers can expect the same level of mobility that they enjoy with telecom services. In a broader sense data portability includes moving data between applications within the same provider. Since more and more organizations use cloud to store data the need for data portability has evolved to be an important component of cloud service. However, the cloud service providers have not developed a standard process for facilitating data portability. One reason for this is the lack of cloud standards that gives the cloud service provider the freedom to use their own proprietary format to store data in the cloud. The APIs that the cloud service provider supports facilitate the use of popular applications such as Outlook email, Office 365 and CRM applications from Salesforce. The data portability involves having a process to move such data from one service provider to another and be able to use the same applications. This aspect of cloud service is still evolving due to lack of global standards. However, the Open Stack, Cloud Security Alliance, IEEE and similar consortium efforts are steps in the right direction for data portability (IEEE 2014; Cloud Security Alliance 2014).

Data portability related problems are multi-faceted. The most expected data portability aspect relates to the ability of the customer to switch providers and have their data transferred to the new provider quickly. A recent European Union Parliament resolution calls for "complete data and service portability, and a high degree of interoperability between cloud services, in order to increase rather than limit competitiveness" (EU 2013). Thus, the importance of data portability aids not only the customer but also increases competitiveness. Data portability also means the ability to have access to data between multiple applications based on central cloud storage. This means that a social media application would need access to contact information for friends from an email system. Another aspect of data portability involves data integration. All users want data integrity and so a data updated in one system should automatically make available the updates to other systems that have used the data in the past. At present there are several unknowns with data portability. When a cloud customer decides to port data then the cloud provider should make known in what standard format that data will be ported, how long will the data be available for porting, at what data rate and bandwidth the customer can expect to port the data? The question of cost of data porting should also be specified in an upfront contract rather than at the time of data porting. Often the cloud service providers in their standard contract include a "data hostage" clause which means that the data will be released only upon all payments due the provider. This is not disputed when the porting takes place at the end of a contract. However, when the customer terminates the contract for cause in their opinion, then the cloud provider may not agree and want an early termination penalty paid, which would be significant. Since access to timely data is critical for businesses this matter could not be left up to litigation. The cloud provider and customer should agree to provisions at the initiation of contract as to how data portability will be handled upon contract termination.

Data portability is well supported at the device level among multiple applications. For example, a user using an email application on a Tablet PC later on has the ability to seamlessly access emails on a mobile device and download attachments. This is data portability. However, the cloud service providers have not matured to the level of application support for multiple devices for data portability. Ordinary consumers are able to integrate their data such as contacts, photos, videos and personal information from multiple social networking sites such as Facebook, Twitter and LinkedIn and use the information on all devices accessing these applications. This data integration capability helps the consumer keep their data portable and maintain its integrity. The social network Facebook allows the users to download their data for backup and offline processing. This helps with data loss prevention. The users can also upload the data to their account for ease of content change. Moreover, the consumer with accounts in Facebook and Gmail will be able to share their contacts across these two different applications easily. The advances in data portability with such individual applications show its importance for customers. Cloud computing, by supporting data portability, can significantly increase the value of cloud service to the customers.

Among cloud services data portability takes on a significant challenge at present. Based on the literature on this topic we recommend the following best practices for data portability. Cloud providers typically include a "data hostage" clause in their contracts to mean that the customer data will not be returned to the customer until the customer pays for all services rendered or pays the contract termination fee. Disputes arise based on the reason for contract termination. Any delay in customer having access to their data is detrimental to their business and so the usual legal process will not meet customer expectations on data availability for porting to another service provider or back in-house. Two possible solutions exist. Federal legislation could force the cloud service provider to return customer data immediately and then pursue their legal claim against the cloud customer. Another option is for the two parties to agree at the outset of the service initiation that an independent arbitrator would rule on the validity of customer claim to data without breach of contract. By design the independent arbitrator decision should be quick in order for the customer to have access to their data for porting elsewhere. There is no federal legislation in the offing and so the arbitrator route is the preferred route for data portability at this time.

For businesses of all sizes their data is the most precious commodity. Business-es need continuous access to their data. Given the possibility of cloud customers switching service from one provider to another there should be a global standard for data portability. There is precedence to learn from the telecommunications in-dustry in supporting customer migration to new service providers. The difficulty in developing a global standard for data portability stems from the fact that too many different types of data are stored—textual data based on applications such as wordprocessing and spreadsheet, audio data such as music, video data. Each of these data exist in multiple formats. In order to support data portability there should be a global standard similar to ISO 27001. There are organizations such as Cloud Security Alliance, Cloud Industry Forum, and Open Data Center Alliance that are developing industry standards for data portability. Since these organizations consist of all the leading global cloud service providers any data portability standard devel-oped by these consortium efforts will lead to a global standard.

Data portability also has an economic impact aspect associated with it. The World Economic Forum in its 2011 report points out, "facilitating system interoperability, enabling users to customize their own cloud solutions across multiple providers, and data portability to ease user fears of vendor lock-in and government fears about lack of competition" are important considerations in the use of cloud computing (World Economic Forum 2011). In order to realize the significant economic impact of 2.3 million new jobs in the top 5 EU economies due to cloud computing and a $ 55 billion revenue globally from public cloud alone, the cloud computing industry has to focus on data portability. Realizing the importance of data portability and interoperability the Open Group, a global consortium of service providers aiming towards common IT standards, has created a report that addresses data portability. The main idea of data portability is the reuse of data components across multiple applications. For example, a business using a Customer Relations Management (CRM) application on the cloud may find that the CRM terms of offering are no longer attractive. In that case it might want to port the data back in-house in order to use a local CRM application (Open Group 2013). It is such scenarios that support the widespread adoption of cloud services. The Open Group's approach is to de-velop standards for data portability that is accepted by the providers and customers. The data portability needs scenario is highly likely because a cloud service provider offering SaaS would switch to a newer version of the product more easily than an individual customer would consider. This could necessitate some changes for the customer in the form of modifying their data for access by the new application. That is when the customer might want to use a different service and port the data out of the service provider.

7.7   Summary

Cloud service customers benefit from cloud services. Cloud customers depend on their contract for the guaranteed level of service. Since cloud services generally tend to use their standard contract available online and executed via the click-through feature, cloud customers do not feel that they have adequate protection afforded by the contract. In this chapter we looked at the hidden aspects of a cloud contract. This analysis showed that customers prefer a customized Service Level Agreement (SLA) whereas the service providers prefer their standard contract. The main conclusion is that the SLAs generally favor the service provider. Since the contract refers to a web content managed by the service provider, the details are changed periodically without much recourse for the customer. The analysis of cloud contracts led to the discussion on the importance of sharing log data with the customers so that they would be able to meet their compliance obligations. The cloud customer concern is usually with respect to security and service availability. It was pointed out that all major cloud service providers offer a 99.9 % service uptime guarantee but try to qualify this guarantee in the form of various exclusions. Even though this level of uptime is quite significant, given the total reliance of thousands of customers worldwide on cloud services, any small outage in service leads to disruption for a very large number of businesses. On the security side, cloud customers are concerned about the access to customer data by the privileged users at the service provider. In this discussion we pointed out the alternatives available that would enhance customer trust based on the level of privileged access users have to their data. The chapter concludes with a detailed analysis of the importance of data portability for customers and how governments and standards bodies are trying to come up with new laws and standards supporting data portability. Since cloud computing is still maturing as a service it will take some more time before global standards are in place that will support full customer mobility within various cloud services, just like people around the world enjoy mobility in telecommunications service.

7.8   Review Questions

1. Explain the importance of a contract for the cloud customer. What three issues impact the cloud customer the most with respect to the contract? Describe each issue.
2. In what ways the cloud service provider is tilting the SLA in their favor? Identify three such scenarios and describe each.
3. Analyze the significant features of SLA from AWS, Windows Azure, Google Apps and Rackspace.
4. How is service uptime significant for a cloud customer? What steps could they take to overcome possible service outage at the service provider?

5. How does the service provider interpret the service uptime guarantee? What options are there to mitigate the risk of service outages?
6. Privileged users at the service provider have the ability to touch customer data. What safeguards can be built to notify customer of such access as well as get customer approval for any modifications to customer VM?
7. Data portability is not supported widely by cloud service providers. What are the reasons for such an action? Explain.
8. How will data portability help cloud customers? Explain the advantages.

References

AWS. (2014). Service health dashboard. http://status.aws.amazon.com/. Accessed 30 Jan 2014.

AWS CloudFront. (2014). Amazon CloudFront. http://aws.amazon.com/cloudfront/. Accessed 5 Feb 2014.

AWS CloudTrail. (2013). Security at scale: Logging in AWS. http://media.amazonwebservices.com/AWS_Security_at_Scale_Logging_in_AWS.pdf. Accessed 1 Feb 2014.

Baset, S. (2012). Cloud SLAs: Present and future. *ACM SIGOPS Operating Systems Review, 46*(2) 57–66.

Bleikertz, S., Kurmus, A., Nagy, Z., & Schunter, M. (2012). Secure cloud maintenance: Protecting workloads against insider attacks, ACM ASIACCS Symposium, Seoul, Korea, May.

CloudSigma. (2014). Service level agreement. http://www.cloudsigma.com/. Accessed 1 Feb 2014.

Cloud Industry Forum. (2011). http://www.cloudindustryforum.org. Accessed 30 Jan 2014.

Cloud Security Alliance. (2014). http://www.cloudsecurityalliance.org. Accessed 2 Feb 2014.

Cloud UK. (2011). Contracting cloud services: A guide to best practices. http://www.cloudindustryforum.org/downloads/whitepapers/cif-white-paper-1-2011-cloud-uk-adoption-and-trends.pdf. Accessed 30 Jan 2014.

CSA. (2011). Cloud security alliance security standards. http://www.cloudsecurityalliance.org. Accessed 30 Jan 2014.

DLA Piper. (2012). The European technology index. http://www.dlapiper.com/files/Uploads/Documents/DLA_Piper_European_Technology_Index.pdf. Accessed 30 Jan 2014.

EU. (2013). Unleashing the potential of cloud computing in Europe. http://www.europarl.europa.eu/sides/getDoc.do?type=TA&language=EN&reference=P7-TA-2013-0535. Accessed 3 Feb 2014.

Geyer, A. & McLellan, M. (2011). Strategies for evaluating cloud computing agreements. *Bloomberg Law Reports, 3*(13) 35–37.

Google. (2014). Apps status dashboard. http://www.google.com/appsstatus#hl=en&v=status&ts=1390938261968. Accessed 30 Jan 2014.

Hon, W., Millard, C., & Walden, I. (2012). Negotiating cloud contracts: Looking at clouds from both sides now. *Stanford Technology Law Review, 16*(1), 79–129.

IEEE. (2014). Cloud interoperability standard P2301. https://standards.ieee.org/develop/project/2301.html. Accessed 30 Jan 2014.

JSON. (2014). JavaScript object notation. http://www.json.org/. Accessed 1 Feb 2014.

Microsoft. (2014). Windows Azure service dashboard. http://www.windowsazure.com/en-us/support/service-dashboard/. Accessed 30 Jan 2014.

MIT. (2014). Cloud data. http://wiki.scratch.mit.edu/wiki/Cloud_Data. Accessed 1 Feb 2014.

Nimbits. (2014). Distributed cloud. http://www.nimbits.com/. Accessed 1 Feb 2014.

ODCA. (2011). Developing cloud-capable applications. http://www.opendatacenteralliance.org/docs/DevCloudCapApp.pdf. Accessed 30 Jan 2014.

Open Group. (2013). Cloud computing portability and interoperability. https://www2.opengroup.
org/ogsys/catalog/G135. Accessed 3 Feb 2014.

Ponemon Institute. (May 2011). Security of cloud computing providers study. Ponemon Institute.

Rackspace. (2014). System status. https://status.rackspace.com/. Accessed 30 Jan 2014.

Safe Harbor-EU. (2000). US-European safe harbor agreement. http://export.gov/safeharbor/eu/
index.asp. Accessed 30 Jan 2014.

Safe Harbor-Swiss. (2008). US-Swiss safe harbor agreement. http://export.gov/safeharbor/swiss/
eg_main_018519.asp. Accessed 30 Jan 2014.

Salesforce. (2014). Analytics. http://www.salesforce.com/crm/customer-service-support/service-
analytics/. Accessed 30 Jan 2014.

Sundareswaran, S., Squicciarini, A., & Lin, D. (2012). Ensuring distributed accountability for data
sharing in the cloud. *IEEE Transactions on Dependable and Secure Computing, 9*(4), 556–568.

World Economic Forum. (2011). Advancing cloud computing: What to do now? http://www3.
weforum.org/docs/WEF_IT_AdvancedCloudComputing_Report_2011.pdf. Accessed 3 Feb
2014.

Index

S. Srinivasan, *Cloud Computing Basics*, SpringerBriefs in Electrical and
Computer Engineering, DOI 10.1007/978-1-4614-7699-3,
© Springer Science+Business Media New York 2014

Printed by Publishers' Graphics LLC
DBT140515.23.35.1